NATURAL RESOURCES AND GLOBALIZATION: PETROLEUM AND VENEZUELA AS EXAMPLES

Collection « Inter-National »
dirigée par Denis Rolland, Joëlle Chassin
Françoise Dekowski et Marie-Hélène Touzalin

Cette collection a pour vocation de présenter les études les plus récentes sur les institutions, les politiques publiques et les forces politiques et culturelles à l'œuvre aujourd'hui. Au croisement des disciplines juridiques, des sciences politiques, des relations internationales, de l'histoire et de l'anthropologie, elle se propose, dans une perspective pluridisciplinaire, d'éclairer les enjeux de la scène mondiale et européenne.

Dernières parutions

Eduard TOKARSKIY, *La Russie au conseil de sécurité de l'ONU (1991-2017)*, 2024.
Giacomo MANTOVAN, *Les dieux manqués de l'Eelam, Militance, défaite et exil des combattants des Tigres Tamouls (Sri Lanka)*, 2024.
Vladimir POL, *Les relations Union européenne-Inde. Vingt ans de partenariat stratégique*, 2024.
Carole GRIMAUD, *Les étudiants face à la guerre russe en Ukraine. Perception et interprétation des informations*, 2024.
Fátima SA E MELO FERREIRA, Lúcia Maria BASTOS PEREIRA DAS NEVES et Joëlle CHASSIN (*coord.*), *Langages de l'identité et de la différence dans le monde ibéro-américain (1770-1890)*, 2024.
Sergio MOLINA GARCIA, *La relation franco-espagnole lors de la convergence socialiste (1982-1995)*, 2024.
Leila BOUSSAID, *Migrations et droit international*, 2024.
Georges LABRECQUE, *Les frontières maritimes en Afrique*, 2024.
Darío Varela FERNANDEZ, *Charles Lesca (1887-1949) au service de l'hispanisme, du fascisme et de la collaboration*, 2024.
Hervé BAUDU, *Les routes maritimes arctiques. Enjeux économiques et géopolitiques*, édition enrichie, 2024.
Nadji RAHMANIA (dir.), *La Roumanie dans l'Espace Schengen : réalité ou utopie ?*, 2024.
Bernard MOMMER, *Ressources naturelles et mondialisation : le pétrole et le Venezuela*, 2023.
Frans KRAJCBERG et Claude MOLLARD, *Le cri de la terre. De manifeste en manifeste,* 2023.
Odile DANIEL-TRAVAILLARD, *Mémoire d'Albanie. De la dictature stalinienne à la transition démocratique*, 2023.
Michel ROZE, *Dictionnaire raisonné de l'Europe. 200 mots-clés pour comprendre l'Europe*, 2023.
Edison FERREIRA DE MACÊDO, *Éducation et précarité. Les accompagnant(e)s scolaires des élèves handicapés. Brésil et France*, 2023.

Bernard Mommer

NATURAL RESOURCES AND GLOBALIZATION: PETROLEUM AND VENEZUELA AS EXAMPLES

Prologue by Arturo Sosa A.

L'Harmattan

5-7, rue de l'École-Polytechnique, 75005 Paris

http://www.editions-harmattan.fr

ISBN : 978-2-336-49702-0
EAN : 9782336497020

Publication in Spanish:
Recursos naturales y globalización: petróleo y Venezuela como ejemplos, Universidad Católica Andrés Bello, 2nd ed., Caracas, 2023.

Publication in French:
Ressources naturelles et mondialisation : le pétrole et le Venezuela, Harmattan, Paris, 2023.

Translation into English by Juan Carlos Boué and Daniel Hellinger.

Thoroughly reviewed by the author in consideration of the US-American English reading public.

From the standpoint of a higher economic form of society, private ownership of the globe by single individuals will appear quite as absurd as private ownership of one man by another. Even a whole society, a nation, or even all simultaneously existing societies taken together, are not the owners of the globe. They are only its possessors, its usufructuaries, and, like *boni patres familias,* they must hand it down to succeeding generations in an improved condition.

Karl Marx, *Capital,* Vol. III, Chapter 46.

Acknowledgements

I arrived to Caracas in June 1970, as a professor of mathematics, and immediately became fascinated by the encounter with the phenomenon of a petroleum exporting country which benefited from a sizable international ground rent. I was soon able to ascertain, in both the national and international literature, the lack of a systematic analysis focused on international ground rent. I then resolved to fill this gap and, without jettisoning my background as a mathematician, I became a social scientist. The results of my research from that time were incorporated in my book *Die Ölfrage* (The Oil Question) (1983). However, I was still in no position to assess properly either the nationalization of the petroleum industry which took place during the decade of the 1970s in the member States of the Organization of the Petroleum Exporting Countries (OPEC), or their undeniable and extraordinary importance, at both a national and an international level.

In 1976 I met Asdrúbal Baptista at the University of Los Andes in Merida (Venezuela). He was then in the midst of his development as a political economist, rooted in the analysis of Adam Smith's *An Inquiry into the Nature and Causes of the Wealth of* Nations (1776); for my part, I was rooted in the analysis of Karl Marx's *Das Kapital* (Capital), published between 1867 and 1894. A close collaboration arose between us, which resulted, in 1981, in our paper *La Economía Política de Venezuela*, a first outline of a research agenda. This paper already contained the concept of *rent capitalism*; that is to say, the development of capitalism in underdeveloped countries which benefit from a significant international ground rent.

Asdrúbal Baptista also introduced me in Caracas to the Gumilla Centre, a foundation of the Society of Jesus. Between 1989 and 1991, crucial years in the history of Venezuela as an oil country, I participated in a fortnightly seminar organized by this Centre, with participants drawn from a broad spectrum of Venezuela's national life. Among them stood out Ramón Espinasa, later to become chief economist of Petróleos de Venezuela, S.A. (PDVSA); Alí Rodríguez Araque, later to be Minister of Energy and Mines, Secretary General of OPEC and President of PDVSA; and Arturo Sosa, currently the Superior General of the Society of Jesus.

The discussions were exceptionally rich. Nevertheless, we were unable to conclude, as to where the process then under way might be taking us. I therefore sought for the answer in praxis. In January 1991, thanks to the support of Ramón Espinasa, I joined PDVSA's Strategic Planning Unit. But I soon became convinced that the *Apertura Petrolera* (Oil Opening) policy that

PDVSA was then promoting was going down a path which I considered unacceptable.

In 1995, working at the *Oxford Institute for Energy Studies*, I met Juan Carlos Boué, a Mexican economist focused on public administration issues. I benefitted greatly from his extraordinary knowledge about international petroleum politics, and in particular about PDVSA, which he generously shared with me. It allowed me to reach a better understanding about the Oil Opening policy then under way. However, although I now understood the path which should *not* be followed, I was only able to take the analysis forward following its development, but without a historical vision of my own, as comes across in my book *Global Oil and the Nation State* (2002).

In 1999, with the appointment of Alí Rodríguez Araque as Minister of Energy and Mines of Venezuela, I made a second attempt to find there, by means of praxis, the answers that I was seeking, and became his advisor. PDVSA's Oil Opening policy was now questioned. In any event, I could continue to rely on the support of Juan Carlos Boué and, on his recommendation, I contacted the New York law firm *Curtis, Mallet-Prevost, Colt & Mosle LLP*, in an effort to understand the legal and juridical dimensions, both national and international, of the petroleum question.

In January 2005, the Minister of Energy and Mines of Venezuela, Rafael Ramírez Carreño, appointed me as his Vice-Minister for Hydrocarbons and, in August 2008, as Venezuela's OPEC Governor. The doors of praxis were thus thrown wide open, a fact which finally allowed me to write the present book, and I want to express my special thanks to Rafael Ramírez.

In this new setting, besides the continued backing of Juan Carlos Boué, I was also able to draw upon the support of the law firm *Curtis et al*. Beyond its advisory role regarding my activities as vice-minister, this law firm would take on the representation of PDVSA as well as that of the Bolivarian Republic of Venezuela, in the international arbitrations subsequently started by some oil companies. I would like to thank, in particular, George Kahale III, Benard Preziosi and Miriam Harwood; it is to their work and dedication that I owe all of my knowledge about international law and, in particular, about the development of international arbitration. As a matter of fact, I was finally able to understand the petroleum question as part of the wider trend whereby natural resources in general have been subordinated to the process of globalization of capitalism.

For his part, Asdrúbal Baptista devoted himself for many years to study and produce the statistics of Venezuelan rent capitalism and, thanks to his efforts, I was able to include its development in the present book. But when I finished the chapter in question and sent it off to him in order to benefit, as had always been the case, from his critical opinions, in lieu of an answer I got the shocking news of his sudden death. Beyond the loss to Venezuela of its most renowned political economist, at a personal level I lost an exceptional friend.

In 2018, I began to work on the present book. In my effort to place the problem of natural resources within the wider context of globalization, I was always encouraged by Robert von Hackwitz. I met him first in Tübingen in 1968, and we always shared an interest in placing the political issues then under discussion within their wider context, and not settle for limited interpretations of particular cases. An insatiable reader, his discussions on the topic of the present book were very important and stimulating for me.

My wife Irama, who was involved in discussions and research on the petroleum question since 1971, later joined me, as a historian and political scientist, in this journey, taking up the most varied positions as circumstances demanded, as a researcher and also as an administrator. Her critiques and contributions constitute an integral part of this book. She, in turn, brought on board a Venezuelan researcher who is currently engaged in preparing his post-graduate thesis in Paris, Carlos Luis García. His practical help, reviewing my drafts time and time again, has been invaluable.

I would like to thank Estela Aganchul for her work in revising and editing this book with a view to its publication. I first met her in 1976, and she was already working in the field of publishing. However, as an Argentinian political refugee in Venezuela, she always read and still reads my works with a great political interest, having her own country in mind, a major exporter of agricultural products which has historically benefitted from a substantial international ground rent. Her critical observations, from such a different perspective, were always very pertinent.

Lastly, as is usual, I would like to put on record that, whatever the merits of the persons mentioned or quoted in this book, all errors to be found in it are my sole responsibility.

Bernard Mommer
Vienna, June 2021

Prologue

Spending time with the future is a necessity of the period humanity is experiencing, if we really do wish to give future generations a life. Reading this book makes us aware of the challenge we face as persons, and of the enormous responsibility that peoples, cultures, nations and international structures bear in terms of the creation of the conditions necessary to guarantee that upcoming generations will inherit a relationship that is reconciled with the environment.

Spending time with the future leads us to perceive the urgency of giving priority to the policies and actions which will make it possible to bequeath to the forthcoming generations a better planet Earth. Spending time with the future means assimilating the intercultural and intergenerational character of the decisions of the present, if one wants to advance towards the humanization of the relationships between persons, peoples and the environment, making possible a dignified life for all human beings.

Starting from petroleum, that non-renewable natural resource, which has been the main energy source for capitalist development for more than a century as well as the fountainhead of the dream of a better life for entire nations, Bernard Mommer studies the relationship between globalization as an intrinsic necessity of capitalism, and guiding principle for the production of the goods that our present civilization requires, and the limited natural resources without which life on Earth would not be possible.

Spending time with the future requires remembering the past in order to understand the present and the path that has taken us to it. Remembering means not only recalling events or historical stages, but also understanding the guiding thread that allows us to explain the decisions taken by humanity in those times, and their impact on the structure of the present, the sole point of departure towards the future. This book calls to memory the experience of oil countries, with special emphasis on the history of Venezuela, where I was born and developed personally, spiritually and intellectually, and where I have spent the major part of my life. It is a book that challenges collective beliefs generated by mental images insisting on associating petroleum with a collective economic well-being automatically arising by its possession as a natural resource. It challenges these collective beliefs which have, indeed, limited our capacity to understand our own selves.

Spending time with the future makes it possible to get out of the limits of one's own vision. Reading Bernard Mommer's complex and detailed analysis of Venezuela as an oil country not only allowed me to re-trace the thread of a history that I, in good measure, have lived through and have also striven to understand in depth, but it also offered me new insights that helped me to leave behind well-trodden but trite paths. We Venezuelans have not found it easy to get to know our history in a dispassionate fashion, going through its nooks and crannies in order to understand it, and not to justify or condemn it. Not opening ourselves to a serene memory of the past makes it difficult to understand the present and, consequently, to propose a coherent course of action towards a future that presupposes fairer social relations and a better environment, being taken care of as the authentic common home of a people that takes collective decisions looking at the good of the whole.

The relationship of Venezuelans with the natural resource, with petroleum, has been marked by the social and political conflicts of the 20th century, intimately connected with the vagaries of the postcolonial process. From the end of the long war of emancipation (1824) and up until the emergence of petroleum as the main source of public resources (1925), enormous difficulties were encountered to unite society and to find a productive support for the development of the country. The positivist ideas implanted in the prevailing mode of thought the need for a "Democratic Caesar", a guarantee of the "order" required for the "progress" of the nation. They justified the political dictatorship which, during the first decades of the 20th century, organized the Nation State and the Armed Forces that served as its support. The transition of Venezuela towards an oil country, on the basis of a concessionary regime, takes place in a country weakly integrated territorially, without technological resources, and with a sparse, dispersed, barely educated population, politically subjugated to a dictatorial regime.

Thereafter, the petroleum question would become associated with ideological debate and political struggle. The debate is about the model of the "modern" society to be built, while the struggle is about creating the conditions that would make it possible. The tensions between dictatorship and democracy, military and civilian rule, capitalism and socialism, neo-colonial dependence and national autonomy, etc., are present in the Venezuelan social process from the beginning of the 20th century. No dimension of Venezuelan life would thenceforth be able to disregard its relation with this vital natural resource. Venezuela turns into an oil country because it organizes itself as a State-led rent capitalism. And, in addition, because politics, social relationships and culture were to be conditioned by the central role of oil in the life of the nation.

From this formative process as a society dependent on petroleum rent arises a rentier culture that becomes the mirror in which the population looked upon itself and determines modes of behavior in the economic and political spheres. It is rentier culture that is the foundation of the collective conviction of being a rich country, now and forever, by virtue of there being abundant petroleum in the subsoil. The view of petroleum as a natural resource that would quickly disappear became one of perceiving it as a source of permanent rent that would secure the future on a basis of certainty. Political struggle became focused on seeking to control the government of a State which was the collector of the rent and also responsible for molding society through its distribution.

The rentier culture became a way of perception and a collective attitude which would make an uphill struggle out of both spending time with the future and to remembering the historic process that leads to the present. The illusion of "Great Venezuela" as a fruit of the end of the concessionary regime during the last third of the 20th century, or of "Venezuela, a World Power" at the beginning of the twenty-first Century, took hold of the collective imagination with such force that the memory necessary to plot the route to the future of a better earth became lost.

A paradox present in an oil country like Venezuela is the relationship between rentier culture and democracy. The model country advanced by the majority of the social forces and political organizations born at the outset of the modernization process, enabled by petroleum income, claims to be ruled by political democracy. The economic, social and political conditions of a postcolonial and pre-modern country are a compromising departure point for a society seeking to organize itself democratically. The emergence of petroleum rent as a lever for the modernization process makes it easier to advance in certain areas, such as infrastructure, which could then support the development of economic and social activity. However, the fact that the State would be the sole collector and distributor of petroleum rent, the most important source of income in the nation, outstripping any other form of private or public income by a long distance, is something that conditions both the style of economic growth as well as the possibility to set up democratic relations.

It is worth remembering the fragility of the state at the moment when the stream of petroleum incomes occurs. What exists then is a Government rather than a State. The Executive Power, in the hands of a dictator supported by the Armed Forces and a well-structured network of selective repression, is the authority where political decisions are taken. The other public powers, constituents of a modern democratic polity, are subordinated to the Executive. The absence of a true separation of Powers makes it impossible, in this context, to conceive a Legislative Power that would represent the national interests, elected by a civil society to which it would be accountable, and issue the laws which would determine the route of government at all levels, approve the budgets that would both govern the use of the resources due to petroleum rent and would demand accountability in the management of such resources. Even more remote is the possibility of an independent Judiciary which would serve as a balance for the other powers. The reality was, therefore, the existence of an institutionally weak State with a Government that may dispose, at its discretion, of increasing petroleum incomes.

At the same time, the process of building the popular subject of democracy is a complex task that confronts adverse conditions. To approach the democratic ideal as a "government of the people", presupposes that such a people exist as a grouping of politically conscious citizens, who are organized for participating systematically in defining and implementing the Common Good of society. An oil country where the government receives and distributes the rent, lacking both an institutional framework to foster the public consciousness of the citizenry and the organization of the people in order to participate in public life, faces too many obstacles for growing as a democratic nation.

Democracy needs powers that arise from the civic will of the people, a set of solid public institutions, with checks and balances that maintain the equilibrium among them; Armed and Police Forces which are subordinate to civilian rule; a civil society that is able to produce what is necessary to make a livelihood; and a State oriented towards the attainment of the Common Good, that is dependent on the citizenry which it serves and which holds it to account. And whereas it is only fair to recognize that major efforts were undertaken in Venezuela, an oil country, to sow and cultivate democracy, and that such efforts bore significant fruit, it is also necessary consciously to acknowledge the fragility of the results obtained and how far away we still are in terms of the necessary conditions for a mature democracy. Spending time with the future is, without doubt, a way of encouraging democratic efforts in oil countries and in a world in which globalization may asphyxiate them.

Among the virtues of this book is that it offers a detailed and complex analysis of Venezuela as an oil country but does not stop there. It offers the Venezuelan case as a source of illumination for the relationship between capitalism, globalization and natural resources in the current socio-historical structure of the world, a topic without which it is not possible to penetrate the recesses of what a better future for humanity could be. Any glance at the global panorama quickly runs into the relationship between the way in which globalizing capitalism exploits natural resources, and the degradation of the environment and armed, social, or political conflicts in those regions of the world where such resources are found in abundance together with poverty, with all of its consequences.

The aspirations for social justice, a dignified living for the whole population and the possibility of political participation in stable democratic systems, succumb before the globalizing voracity of the exploitation of natural resources. What is more, millions of people are forced to abandon their homes, birthplaces, families or social relations because of this mode of exploitation of natural resources. A fundamental part of the rising involuntary tide of migration all across the world is linked to the current relationship between capitalism, globalization, and natural resources.

To spend time with the future is an invitation to make a serious commitment to the transformation of the relationship of humanity with natural resources, which supposes a radical change in the relationships between human beings, peoples. and cultures. A world in which cultural diversity would be considered a human wealth on a par with bio-diversity in nature would be an invaluable legacy to new generations, as opposed to bequeathing them the results of a globalization that crushes cultural diversity. Interculturalism is one of the ingredients for a better future, together with political democracy, international relations which guarantee the Common Good for the whole planet, social justice, and modes of production and distribution of the goods of civilization which are in equilibrium with the environment, and which offer the opportunities of life to every person.

To benefit from this book requires making an intellectual effort motivated by a desire to deepen one's understanding of an important subject matter that casts light on central aspects of the current social structures of the world. It is a hard read, that makes no facile intellectual or literary concessions. It involves legal and political economy concepts that demand a precise understanding. One benefits from this book to the extent that reading it renews the desire to make an effective contribution to improving the legacy that the generation which we belong to intends to leave to subsequent ones.

A good part of this book was written during the unprecedented situation experienced throughout the world as a result of the SARS-CoV-2 pandemic. A situation that for many became an invitation to spend time with the future. Reading this book in such a context, under the sign of the pandemic and its consequences across all the walks of life, helps to reinforce the necessity of living this historical juncture as an opportunity to dream of a different world, to introduce changes to our lives that may bring it forth and to make of our lives a contribution to that better world which we desire for the new generations.

Arturo Sosa A.
April 2021

Introduction

The title of this book highlights a central topic the importance of natural resources in the process of globalization of capitalism. In its beginnings, capitalist development revolved around them because this mode of production arose in the midst of European political structures dominated by a landholding class; whereas the new structures were characterized by the domination of a capitalist class. In this process of transformation, the latter had to subordinate the former.

Ideologically, this process of subordination was reflected in the evolution of modern economic thought. Its starting point was the sphere of production and not that of trade, as had been the case until then, and its evolution can be summed up in three successive schools, within a period extending over a century and a half.

The first of these was the French school of *Physiocracy*. For the physiocrats, the foundations of an economy were two factors of production: *labor and land*. Labor was only reproducing itself, and all surplus value originated from land. Physiocracy was followed by the British school of *Political Economy*, for which the factors of production were three: *labor, capital and land*. Labor was now the source of all value, and no longer land. Nevertheless, the salary of a worker only reflected the cost of his reproduction, whereas his surplus labor – or surplus value – was appropriated by the capitalist who, in turn, would have to share it, in the primary sector, with the landholder. Capital had already interposed itself in this sector, in the form of lessees, between labor and land. Hence, surplus value was to be divided between profits and ground rent.

However, with the development of *private capitalist property*, there also developed private capitalist *landed* property, as landholders got the opportunity to become capitalists by selling their lands, and capitalists were offered the opportunity to become landholders by buying them. This process went hand in hand with the ejection of the serfs from the land, throwing them into the labor market as free workers. Ultimately, through the selling and buying of rights over land, the historical line of demarcation between the landholding and capitalist classes was gradually erased, as well as the distinction between the concepts of *ground rent* and *profit*. Land thus became, and appeared as, an investment option, among many others.

Thus, in modern-day *Economic Sciences*, the factors of production once again became reduced to two – now *labor* and *capital* – and the factor of production *land* disappeared from the scene, to be absorbed by the factor of production *capital*. Natural resources therefore became invisible, as they were definitively subordinated to capital. This process is analyzed in the First Part of this book, *National Development of Capitalism and Landed Property*.

National Development

The subordination of land to market forces and, consequently, the dismantling of traditional structures which stood in their way, brought about the accelerated development of productive forces. However, even though landed property at a given moment might have been accommodated accordingly, there remained the possibility that, with the subsequent evolution of productive forces, the existing structure of landed property could once again become an obstacle to capital.

Therefore, even though landed property – as far as the surface was concerned – generally took the form of *private* landed property, it nonetheless remained subject to the eminent domain rights of the state over all the natural resources that make up a national territory. However, regarding mineral deposits which might exist in the subsoil, the situation was quite different. With the transition from feudalism to capitalism, the majority of European countries maintained the historical division between rights over the surface, on the one hand, and rights over mineral deposits, on the other. Private landed property owners of the surface were obliged to grant mining capital access to their areas for their exploration and eventual exploitation, and the state conferred the corresponding mineral rights by means of *concessions*. The exploration for, and exploitation of, minerals were activities that could only be carried out with the permission of the state, and the concessionaires were obligated to indemnify the surface owners for any damages that mining activities could cause to them, but they had no right to a ground rent.

In Great Britain, contrary to what generally happened elsewhere throughout the Continent, the development of private landed property would also extend, albeit with some exceptions, to mineral deposits as, for instance, coal, which was of extraordinary importance in the development of British capitalism. In order to justify the private property rights over mineral deposits – a compromise which required some justification – Economic Sciences took refuge in the conception of seeing these as stores of 'valuable commodities'. As a consequence, a *part* of the rents and royalties would reflect the *intrinsic value* of the mineral, and would therefore not constitute a ground rent. The latter, in turn, was thereby reduced to the concept of a *differential rent*, as the expression of an exceptional natural productivity of the better quality lands.

However, with the passage of centuries, the ever-increasing depths of coal mines entailed that the division of the surface into private parcels overlying one same deposit became ever more inapposite in terms of achieving a rational form of exploitation. Finally, in 1938, the British government nationalized the natural resource. Private landed property owners were fully indemnified for the value of their rents and royalties, as the benefit of nationalization was supposed to be the subsequent development of productivity, by means of the unified management of the mining companies exploiting the same deposit.

Moreover, with nationalization, the concept of coal deposits as stores of 'valuable commodities' simply vanished. Once coal resources became part of the public domain, it was taken as read that coal mining would simply generate *taxes*.

In the case of petroleum in the United States of America – of an extraordinary historical importance in the development of capitalism in that country – private mineral property was imposed by the settlers in their conquest of the national territory, against the incipient state and federal authorities. Then, however, the rights of surface owners were reformed, and made to fit with the development of productivity in the industry, albeit without denying them the right to a stipulated ground rent fixed at a certain level. The states and the Federal government managed leases in public lands and waters in a similar fashion.

The capitalist lessees, as part of the capitalist *class*, unified and dominant, got their way. Surface owners, in contrast, represented a very important set of individuals which were generally devoted to the most diverse agricultural activities and definitely did not conform a landholding *class*. At present the USA is the only oil producing country where the mineral deposits are still part of private landed property.

Hence, in the collective imaginary of the American people, petroleum deposits are also associated to stores of 'valuable commodities'. Moreover, when income taxation was created in the 20th century, the law even admitted a corresponding entry.

Internationalization

In their drive to subordinate land to capital at a national level, the developed countries also turned to lands throughout the rest of the world in order to source agricultural, livestock and mineral products. There arose then groups of underdeveloped countries which exported primary products and imported manufactured ones.

This book refers to the *example of petroleum*. This choice is due to the fact that, as a source of international ground rent, petroleum, on its own, exceeds in importance all other raw materials put together.

Regarding petroleum, the USA always occupied a special position among the developed capitalist countries, not only by virtue of private mineral property, but also for being – until 1947 – the only developed petroleum exporting country. Nevertheless, the USA, definitively a capitalist state, was fully on board with the twin imperatives of minimizing the incidence of private mineral property rights on the industry, on the one hand, and of keeping ground rent under check, on the other.

All the other developed capitalist countries understood that petroleum resources in the public domain simply constituted a free gift of nature. In particular, this was also their position with regard to the underdeveloped petroleum exporting countries. However, these exporting countries would come to understand it, specifically, as a free gift of nature for the *nation* and, in consequence, as a *national property* that should serve as a source of an *international* ground rent.

There developed, then, a confrontation between these two groups. The developed capitalist countries questioned the legitimacy – and even the legality – of that conception, and refused to accept that the sovereign rights of a country could be put in the service of international ground rent. However, this happened to be precisely what the underdeveloped petroleum countries aspired to do, thereby transforming themselves into *landlord nation states* and, indeed, *sovereign* landlord nation states.

This confrontation developed within the boundaries of the international concessionary regime and culminated in 1973 with an overwhelming victory for the petroleum exporting countries. We analyze this regime in the Second Part of the book, under the title *International Concessionary Regime*.

Globalization and Fiscal Regimes

However, the developed capitalist countries would never turn their backs on what is part of the essence of capitalism, and therefore of its globalization: namely, the subjugation of natural resources. In order to further their interests, it was necessary to create a new global petroleum regime. The developed capitalist countries now insisted in eliminating from the process of granting exploration and production rights any feature which would have a bearing on marginal production costs. The new fiscal regime would only target differential rents, on the basis of excess profit taxes. Hence, the recovery factor of the deposits would be higher and, with greater output would come lower prices.

Nevertheless, the exploitation of more naturally productive oil fields would continue to generate an extraordinary surplus, due to the imperfection of competition in the primary sector, as their production costs are irrelevant in determining the market price. To remedy such imperfection, there arose the proposition *to oblige* the company exploiting a given field, to extend to the state a *mandatory loan* of sorts, which would become a *contingent liability* of the state. The state would thereafter liquidate this loan by paying, on an annual basis, whatever sums were necessary to ensure that the exploiting company would continue to obtain a reasonable rate of profit, notwithstanding the fact that its production costs would translate into low profits or even losses.

Therefore, windfall profits would then no longer be quantified on an annual basis, but over the lifetime of the field in question. Hence, production would not be abandoned as soon as lifting costs exceeded market prices, but only at that point when the *average* lifting costs, measured over the whole of the lifetime of the field, exceeded market prices. The consequence of such an arrangement would be that the recovery factor would be even higher and prices therefore would be even lower. Furthermore, this would be a *global* regime, because it would minimize windfall profits across the board, irrespective of whether these arose domestically or internationally. *Globalization and Petroleum Fiscal Regimes* is the topic of Third Part.

From a Concessionary to a Cessionary Regime

From Part Four to Seven, we shall consider the various regimes from the perspective of the underdeveloped petroleum exporting countries. Now, these countries were and continue to be very different amongst themselves, quite apart from the fact that, individually, they found themselves in the midst of the most varied and complex international circumstances, so therefore these regimes found expression and developed – culturally, politically, legally and institutionally – in a very specific form in each case. As regards this globalization process, we shall restrict our examination to the example of Venezuela.

The choice of Venezuela is due to the fact that it is the oil country par excellence, both from a quantitative as well as from a qualitative point of view. In no other oil country has the historical record been as complete as in Venezuela; furthermore, this history has not been linked to the complex international circumstances that prevail in the other petroleum countries.

In the Fourth Part, we refer to *Venezuela: The Petroleum Regime*. When the petroleum industry was nationalized in 1975, Venezuela also put an end to the concessionary regime, and the national oil company was placed between the multinational oil companies (historical representatives of the interests of the developed consuming countries), on the one hand, and the landlord state, on the other hand. Hence, a *concessionary regime*, managed by the Ministry with a view of the natural resource as a source of international ground rent, gave way to a *cessionary regime*, managed by the national oil company with a view of petroleum as a productive activity. Thus, the petroleum policy of a landlord state morphed into a policy of a shareholder state.

Rent Capitalism

However, to understand the consequences of this change of the petroleum regime, it is necessary to grasp first the characteristic problems of the development of a national economy based upon an international ground rent. Given its significant weight in relation to the size of the national economy, the development of Venezuelan capitalism took the form of what we have called *rent capitalism*, which we will analyze in Part Five, *Venezuela: Rent Capitalism*.

After being remarkably successful throughout the first decades, during which the country developed into an oil country, it became increasingly difficult to maintain high levels of growth, even though the international ground rent continued to climb. By channeling this international ground rent to the national economy, consumption levels – through imports – would always expand, but the development of productive forces started to lag behind. This was essentially due to the elementary fact that the capital absorption capacity in any economy is determined and, ultimately, limited, by its productivity. But the international petroleum rent was not related to the productivity of the national economies of the exporting countries.

Nevertheless, when the international petroleum rent expanded by leaps and bounds at the beginning of the 1970s, the Venezuelan government engaged again in a project to accelerate the development of the country, which was supposed to result in a 'Great Venezuela', completely unaware of its limited capital absorption capacity. And the government spent not only the very high international petroleum rent arising at that juncture, but also resorted to the increasing capacity of the country to get indebted, which expanded in lockstep with the new level of the international petroleum rent.

This is a noteworthy fact when one considers that, in 1930, Venezuela had taken advantage of the bonanza fiscal generated by petroleum to liquidate all of its external debt. Subsequently, the foreign concessionaires also opposed any indebtedness of the country, being conscious that they would be the first ones who would have to foot the bill of the foreseeable failure of higher levels of investment. But with nationalization, this was no longer their concern.

As the country got into foreign debt just as prices were at their highest, it would be unable to meet its obligations when prices came down again. Then, Venezuela would have to turn for emergency loans to the international financial institutions, all under the control of the developed capitalist countries. Evidently, these institutions would then be in a position to impose their terms on the country.

Cessionary Regime

Inevitably, the project of a 'Great Venezuela' ended in failure and the success of rent capitalism would be definitely limited to the concessionary period.

In 1989, when Venezuela needed emergency loans, the international financial agencies imposed their conditions, requiring an optimum management of state assets. In the first place, obviously, was petroleum. The argument was that, by reopening the petroleum sector to private investment, resources would be liberated that would enable the country to repay its debt. This policy became known as the Oil Opening (*Apertura Petrolera*) and with its implementation, the international oil companies would come back, however not as concessionaires but as contractors of the national state oil company.

The Oil Opening had the support of the national state oil company, as well as of the national business community in general. Definitively, they all blamed the economic crisis on the incompetence of the state. Never was the connection made that the origin of the problem was to be found in the foreign indebtedness, at a juncture when the country was benefiting from an extraordinarily high international petroleum rent, way beyond the capital absorption capacity of the national economy. In those days Venezuela should have become an international *creditor* country, investing surplus funds *abroad*, instead of a *debtor* country. But now, as a highly indebted country, Venezuela would have to place petroleum at the forefront once again as a magnet for foreign investment.

On the horizon the definitive consequences of nationalization could be glimpsed. The Ministry of yesteryear had been supplanted by the national oil company and, under the auspices of this company, productive activities would once again be privatized, a topic that we develop in Part Six, *Venezuela: Cessionary Petroleum Regime*. In Part Seven, *Venezuela, an Oil Country: Conclusions*, we synthesize the three parts that deal with Venezuela.

Arbitrated Sovereignty

Now, to stabilize this process of privatization and, in particular, to guarantee that it would be irreversible, it was necessary that the country renounce its most elementary sovereign rights over the national territory, and all the natural resources found therein. Thus, when the multinational oil companies returned in the role of contractors of the national oil company, PDVSA guaranteed them certain profit levels, *against the tax sovereignty of the state*, by committing to pay compensation determined in accordance with an agreed formula in the event that the state exercised its sovereign tax powers. Any disputes in connection with these commitments would be subject to international arbitration before the *International Chamber of Commerce* in Paris.

At the same time, a process got underway which was aimed at the state itself assuming the corresponding guarantees. Venezuela ratified then its first bilateral investment treaty, the Agreement on Encouragement and Reciprocal Protection of Investments between the Kingdom of the Netherlands and the Republic of Venezuela (*Convenio para el Estímulo y Protección Recíproca de las Inversiones entre la República de Venezuela y el Reino Unido de los Países Bajos*). Pursuant to this agreement, Dutch investors in Venezuela – and Venezuelan investors in The Netherlands – were granted an irrevocable right to take investment disputes to arbitration before the *International Centre for Settlement of Investment Disputes* created by the World Bank. Among the investments protected by this agreement was included the right of access to the exploration and exploitation of mineral natural resources.

Historically, then, *arbitrated sovereignty* developed against petroleum exporting countries, although the ultimate aim was to subject the sovereignty of all the natural resources exporting countries of the Third World to similar agreements, promoted by international capital and the governments of developed countries. After the collapse of the Soviet Union, the former communist countries were also subject to these kinds of agreements.

The aim was to subject all of these countries, by means of international arbitration, to a global legal regime which counted – and still counts – with the support of the developed capitalist countries. These countries form part of the network of international treaties and are legally committed to enforcing arbitral awards within their respective territories, against the assets held there by liable state parties.

In other words, the Supreme Courts, not only of Third World countries and former communist countries but also of developed capitalist countries, were thereby subordinated to these international arbitrations. For the latter, this situation was of no immediate concern, as the structure was developed under the assumption that it was solely intended to set Third World and ex-communist countries on the right path in terms of global capitalist development. However, they would come to realize that international capital, promoter par excellence of globalization, would likewise seek to prevail against the sovereign rights of developed countries. For example, when these countries implement policies against climate change and in favor of the protection of the environment, they find out that foreign investors had made sure their investments would be protected against such measures. The latter are thus able to claim maximal compensation, through international arbitrations.

International capital has created an international legal structure to protect itself against any policy change involving natural resources, and environmental protection measures in general, which could affect its profitability. The capital importing state, whether developed or under-developed, thereby appears as a partner in a business involving two parties, and *not as a sovereign*. This is because foreign capital supposedly only invested in the country in question on the basis of some *Agreement on Encouragement and Reciprocal Protection of Investments. Arbitrated sovereignty*, and the consequences derived therefrom, is the subject matter of the Eighth, and last, Part of this book.

Part I.

National Development of Capitalism and Landed Property

Modern economic thought is quintessentially European, rooted in the national capitalist development of European economies during the 18th and 19th centuries, above all in the French (Physiocracy) and the British (Political Economy and Economic Sciences) economies.

1. Physiocracy

For Physiocracy, labor in the primary sector was the sole source of *surplus value*, over and above wages, due to the productivity of land, whereas labor in the secondary and tertiary sectors – industrial activities and services, respectively – were seen as unproductive because they were limited to transforming certain goods into others, or simply to putting them in circulation. Thus, land was the only source of surplus value, and landlords, as agents of its exploitation, were considered to be its legitimate beneficiaries (Quesnay, 1758).

Nevertheless, the physiocrats were the pioneers of capitalism, as their stance on taxation reveals. Given that only land generated surplus value, they held that the burden of taxes should fall, first and foremost, on landlords, and not on labor, a category which at that point still encompassed the incipient entrepreneurial class.

2. Political Economy

In Political Economy, capital appeared as a third factor of production which in the primary sector interposed itself as the lessee between landlords and the peasantry. Landlords would no longer act as productive agents, but they would still benefit from a ground rent, as a consideration for granting the capitalist lessee access to their lands.

Labor was now seen as the sole source of value. Surplus value would accrue, in principle, as *profit* to the capitalist as the agent of production. Moreover, the landlords' posture of granting access to land only through lease contracts saw them cast as parasites who obstruct productive activities in the primary sector, and they were blamed for dedicating their ground rents to conspicuous consumption and not to investment, that is the development of productivity. Thus, *the legitimacy of ground rent was questioned on account of its allocation*.

2.1. Adam Smith

Nevertheless, the factors of production, land and capital, were forced to co-exist, sharing the monopoly of ownership of the means of production in the primary sector and, therefore, also the economic surplus – which had to be divided into ground rent of the landlords and profits of the capitalist leaseholders.

In normal capitalist conditions of free competition, typical of the secondary and tertiary sectors, the most productive capital is the one that will end up setting the prevailing market price. In contrast, in primary activities – agriculture, livestock husbandry and mining – nature sets limits to competition, which in turn translates into different levels of economic yield for the better and the worse lands. Consequently, *it is the most productive capital in the worst natural conditions* which determines the price that is to prevail in the market, because better lands are scarce. Therefore, in spite of free competition, there are *enduring* extraordinary surpluses (which, in the long term, are of course subject to technological development and improvements in transport and communication). Hence, free competition will force the potential leaseholders to offer landlords a ground rent equal to these extraordinary surpluses, while landlords will also be minded to accept such offers. Extraordinary surpluses are thereby transformed into ground rent, and not into extraordinary profits, and leaseholders have to be satisfied with achieving no more than the usual rate of profit.

Furthermore, on the basis of his empirical observations of the British economy, Adam Smith contended, at least as far as the production of corn[1] was concerned, that the landlord class was powerful enough to impose a monopoly rent and, ultimately, to impose a market price which exceeded the value of corn. This was a matter of greatest importance, given that corn was the major component in the diet of the working class and, thus, of wages. Any surcharge would not only affect lease holding capitalists and their labor force, but also capital and labor in the secondary and tertiary sectors. All of them would suffer the consequences, namely lower real wages and profit margins. Hence, the whole working and capitalist classes were economically and politically united in their confrontation with the landholding class.

2.2. Count Mirabeau

In 1789, barely thirteen years after the publication of Adam Smith's *Wealth of Nations*, the French Revolution put an end to that monopoly of feudal origins of a landlord class. *The land to the tiller* was the slogan which would set the agenda for the revolutionary agrarian reform. Thus, in 1791, the National Assembly sanctioned private property rights of occupiers cultivating the surface of the land.

Private ownership of the land surface thus became the rule, with a prevalence of smallholdings fully exposed to market competition. However, it remained subject to the eminent domain rights of the state over all of the

---

[1] In British literature, for obvious reasons, 'corn' always refers to wheat or oat, *never to maize*.

natural resources within the national territory. It was up to the state to define how land may be used in general terms, as well as setting diverse limitations on the rights of surface owners as, for example, where homes might be build or not. An example in this regard, which is of particular importance, is the relationship between ownership of the surface and ownership of underground mines.

Indeed, that same year 1791, the National Assembly also decreed that any minerals lying at greater depth would be part of the *public domain*. The debate focused on private land tenure in so far as it constituted a hindrance to the efficient exploitation of mines. The projection of the boundaries of private holdings on the surface down into ever greater depth generated increasingly greater obstacles to the rational exploitation of mineral deposits, as they corresponded less and less to the divisions required for their efficient exploitation.

The right to exploit mines would be granted by concessions, and the corresponding areas would be demarcated irrespective of the boundaries of private plots of land on the surface, and mining was declared to be of public utility. In the words of Count Mirabeau, main spokesman of the Mining Law Bill under discussion, "the declaration of public utility has no other goal, no other motive, than the exploitation of the mines" (Mirabeau, 1792, p. 441)[2]. Surface owners could not deny the concessionaire the right to enter their land in order to access the mines, and neither would they be allowed to demand any compensation for granting such a right of access, over and above the amount of any actual damage that they might suffer as a result of the mining activities. Along the same lines, the state would also not be the owner of the mineral resources, but only their steward or administrator. Therefore, Mirabeau proposed that:

> The national assembly decrees, *as a constitutional article*, that minerals, both of metals and non-metals, as well as bitumens, peats and rock coals and pyrites are at the disposition of the nation, but only in the sense that these substances cannot be exploited other than with its consent [...] (Mirabeau, 1792, p. 491; emphasize in the original).

Minerals would thus constitute a *free property of the state*, and their *exploration and exploitation* would require an authorization from the state; in other words, these activities could only be carried out pursuant to a concession...

---

[2] In this book, unless a statement to the contrary is made, all the quotes in languages other than English were translated by ourselves.

In sum, in the case of minerals, private property rights over the surface had to be restricted to prevent their becoming an obstacle to the exploration and production of minerals:

> The nation has the right to the exploitation of the mines. Therefore, if the mines are not exploited, the nation should procure their exploitation (Mirabeau, 1792, p. 483).

Mining concessions would not have a defined term, and would continue in force for as long as concessionaires were exploiting them, but they would lapse for the failure to do so. More generally:

> The nation has a right to exploitation: but to no more than to them being exploited; so its intervention should not take place whenever this exploitation is basically assured (Mirabeau, 1792, p. 489).

However, when market mechanisms were not able to guarantee the effective assignation of mining rights, state intervention was definitively required.

In principle, all natural resources, the soil and subsoil alike, constituted free gifts of nature. What is commonly called either private or public property only represent different forms of implementing the precept of *the land to the tiller*. Minerals can only be accessed through the surface and the surface owner is under a legal obligation to grant such access. The 1791 Mining Law constitutes the foundation of French legislation on mineral property to the present day.

2.3. David Ricardo

Already in the preface to his work *On the Principles of Political Economy and Taxation,* published in 1817, 41 years after Smith's masterpiece, Ricardo referred to Adam Smith's positions on ground rent as being erroneous. In fact, rebutting Smith was one of the main aims of his book.

2.3.1. Agricultural Rent

First and foremost, Ricardo denied the existence of a monopoly rent in the production of corn, as maintained by Smith:

> The reason [...], why raw produce rises in comparative value, is because more labour is employed in the production of the last portion obtained, and not because a rent is paid to the landlord. The value of corn is regulated by the quantity of labour bestowed on its production on that quality of land, or with that portion of capital, which pays no rent. Corn is not high because a rent is paid, but a rent is paid because corn is high; [...] no reduction would take place in the price of corn, although landlords should forego the whole of their rent (Ricardo, 1817, Chap. 2).

Ricardo's position was based on a *model*. In agriculture, ground rent was typically paid on an *annual* basis. The leaseholder would produce at increasing costs up to the limit where production costs would equalize the market price, and at that point the extraordinary surplus would be equal to the agreed ground rent. The *marginal* product, as economists call it, would not pay any rent at all, but it would be the one which would determine the market price. Hence, all ground rent would be *differential rent*.

That said, Ricardo had nothing to counter Smith's *empirical* observations concerning the existence of a monopoly rent in the production of corn. Ricardo limited himself to building a *model* in which the marginal product would not pay any rent at all.

2.3.2. Mining Rent

Secondly, the most important and typical component of a mining rent was a royalty, a fixed amount per unit of output. A royalty has the virtue, from the standpoint of the landlord, that it is linked to the intensity of exploitation of a mineral deposit, always subject to *depletion*. Certainly marginal mining rent was not zero, and therefore had an incidence on the cost of production of every unit of output produced. However, Ricardo got around this problem by re-defining the concept of ground rent:

> Rent is that portion of the produce of the earth, which is paid to the landlord for the use of *the original and indestructible powers of the soil* (Ricardo, 1817, Chap. 2; emphasize ours).

Hence – taking coal mines as an example – the usual minimum rent would be paid:

> [...] for the value of the coal [...] which can be removed from them, and has no connexion with *the original and indestructible powers of the land* (Ricardo, 1817, Chap. 2; emphasize ours).

For Ricardo, labor would continue to be *a* source of value, but only side by side with *another* such source: land. Mineral deposits would therefore constitute stores of a "*valuable commodity*" (Ricardo, 1817, Chap. 2; emphasize ours). Beyond what was supposed to be the 'value' of minerals in their natural condition – definitively, their *intrinsic value* – the mines would also pay differential rents:

> Mines, as well as land, generally pay a rent to their owner; and this rent, as well as the rent of land, is the effect, and never the cause of the high value of their produce (Ricardo, 1817, Chap. 3).

2.4. *Conclusions*

By holding that all ground rent is a differential rent, Ricardo reduced a contradiction between different social classes to a contradiction which was only relevant on an individual basis. Consequently, there was no justification for a radical reform of land tenure, such as had accompanied the French Revolution, since it would benefit neither the working nor the capitalist class in the form of lower prices. Thus, fiscal reforms targeting ground rent would suffice:

> A tax on rent would affect rent only; it would fall wholly on landlords, and could not be shifted to any class of consumers. The landlord could not raise his rent, because he would leave unaltered the difference between the produce obtained from the least productive land in cultivation, and that obtained from land of every other quality (Ricardo, 1817, Chap. 8*).

Hence, Ricardo confined himself to the proposition that any ground rent (which by definition would have to be a differential rent), should preferably be targeted by general taxation, a proposal which the physiocrats had already put forth. Ricardo's prime concern was to avoid any questioning of private property, even in the case of private landed property.

Finally, it is worth pointing out that, with the development of economic thought, there arose a corresponding vocabulary. Thus, in the economic literature of today, differential rent is usually called *Ricardian rent*, a term that we shall also use. Further, what Adam Smith identified as a monopoly rent, we will call a *proprietorial rent*, because it arises from the private ownership of land as such[3].

3. Economic Sciences

Towards the end of the 19th century, capitalism was already firmly established, and Political Economy was displaced by Economic Sciences of the present. The latter, renouncing any theory of value, would henceforth concentrate solely on prices. Prices, under conditions of what would come to be called 'perfect competition', would converge towards marginal production costs, thereby reducing marginal profit to zero. Obviously, this was what Ricardian theory had already established regarding ground rent, and this theory was incorporated without further ado into Economic Sciences. Definitively, the marginal production cost of land being zero, there was no reason for Economic Sciences to maintain land as a third factor of production. The factors of production were now reduced to two: *labor and capital.*

---

[3] Karl Marx called it *absolute rent* (Marx, 1882, Chap. 45).

Moreover, the sale and purchase of land became an everyday commercial transaction. Hence, for the buyer, land appeared just as some 'investment', and the corresponding ground rent assumed the guise of a 'profit'.

3.1. Mining Rent

As far as the typical mining rent – the royalty – was concerned, Alfred Marshall, with his book *Principles of Economics* (1890), would follow Ricardo:

> A royalty *is not* {emphasize in the original} a rent, though often so called. For, except when mines, quarries, etc., are practically inexhaustible, the excess of their income over their direct outgoings has to be regarded, in part at least, as the price got by the sale of stored-up goods – stored up by nature indeed, but now *treated as private property* {emphasize ours}; and therefore the marginal supply price of minerals includes a royalty in addition to the marginal expenses of working the mine (Marshall, 1890, Book V, Chap. 10).

Note that Marshall considered that the price of these 'goods stored by nature' stemmed from the private ownership of minerals.

Even if capital, once its hold on political power had been consolidated, would deal with the topic of private landed property with utmost caution, so as not to give rise to a critical and potentially revolutionary discourse about private property in general, it would always continue to be the essence of capitalism to subordinate land tenure to a capitalist design. It would always be an *arrangement* in accordance with the circumstances; that is to say, a compromise seen as politically expedient and economically tolerable, and justified by whatever available arguments were most convenient. Nevertheless, capital will always be ready to question and modify such an arrangement, if need be up to the point of putting an end to private landed property in specific cases. In particular, this can happen when landed property becomes an intolerable obstacle to the development of productivity, for example, as a result of the ever-increasing depth of mines.

3.1.1. British Coal: 'Centralization of Royalties'

Historically, British landlords had been the agents of mineral exploitation – in particular, coal – in the same way as they had been in the case of agriculture and livestock. However, after centuries of mining at ever-increasing depths, they ended up turning into mere rentiers vis–à–vis an industry that was becoming ever more complex from a technical point of view, and increasingly capital-intensive. At the same time, mining leases went from having terms of a few years – when all that coal mining involved was shallow excavations – to terms stretching over 21, 42 or even 63 years, on account of deep mining and a structure of investment which required a much longer period for amortization. Nevertheless, reversion continued to be a major obstacle in the path of a continuous flow of investment, because in the bargaining process between landlords and their lessees whenever leases came up for renewal, sunk capital always served as a bargaining chip in the former's endeavors to extract, at the very least, a bonus from the latter. However, in the traditional producing zones, royalties converged to a customary amount of between five to seven shillings per ton, depending on quality differences.

At the same time, the irrationality of private mineral ownership became increasingly obvious. But beyond market forces, the political power of the British landholding class made it impossible to reform it, for example, imposing a legal obligation on neighboring landholders whose properties overlaid the same coal deposit, to collaborate in order to make the joint exploitation of the deposit by their respective lessees possible. Finally, from 1880 onwards, the productivity of British mines began to decline and, after 1913, output also began to drop.

In 1919, in reaction to the poor performance of the coal mining industry during the First World War, a *Liberal* government announced that it would seek the 'nationalization of royalties' and foster regional mergers between mining companies. Then, in 1938, a *Conservative* government actually implemented this program, albeit referring to it by the less contentious name of 'centralization of royalties'. Needless to say, landlords were indemnified in full for their surface rents and the bonuses set forth in the lease contracts, as well as for their royalties. In what follows we shall refer to them as *rents and royalties*, because this nomenclature is usual in the USA. However, we should not lose sight of the fact that a royalty is nothing more than a particular form of ground rent.

The benefit of their *nationalization* was supposed to be an increase in productivity once private mineral property had come to an end. The Second World War broke out a year later and, following the end of the hostilities in Europe in May 1945, the *Labour* Party triumphed in the general election held in July 1945. Thus, the nationalization of the natural resource was followed, in 1946, by the nationalization of the coal industry. In 1994, a *Conservative* government would reverse this last measure, returning the industry to private ownership. However, neither the Conservative Party, nor the Labour Party, nor the Liberal Party, would ever again discuss, or even mention, the possibility of re-privatizing the natural resource. Public ownership of the mineral was to be the definitive, irreversible and unquestionable solution. Nevertheless, political prudence made it advisable not to make too much fuss about a case in which public ownership had been demonstrated to be a superior alternative to private ownership.

3.1.2. American Oil: Conservation

In the USA, the cradle of the international petroleum industry, the private ownership of minerals was imposed by colonists with their relentless advance in the conquest of the national territory. The incipient and remote federal and state authorities were simply too weak to impose, in contraposition to the private ownership of the surface, the public ownership of the minerals that could be found underground. In short, private mineral property finally prevailed also legally.

The first oil leases were signed during the 1860s for limited terms, generally of twenty years. Land surface owners were certainly an important and numerous group, but they were definitely not a social class. Political power was firmly in the hands of capital. Thus, in cases of disputes, the Courts would interpret the legal provisions regarding the private ownership of hydrocarbons, in favor of lessees.

In particular, when the first cases of expiring oil leases arose during the 1880s, the courts backed the right of lessees to dispose freely of their capital and denied the lessors pretension that they were obliged to keep all investment in good condition with a view to reversion. Thus, upon reversion, all that would revert to the lessors would be a pile of scrap or, worse yet, reservoirs in ruins. Therefore, towards the end of the 19th century, the limited term of exploration and production was extended, at one go, for a second, unlimited term, up to the definitive exhaustion of the fields discovered during the first term. Reversion disappeared, and this in turn made possible a continuous flow of investment, as the development of productivity required. On the other hand, in public lands legislation followed the example of the private sector, which, from 1924 onwards, also applied to Native American reservations.

In this new context, landowners would never have an opportunity to adjust their rents and royalties, to which they had agreed upon on the basis of limited knowledge about the potential of specific areas *before* their exploration. Hence, all the additional extraordinary surplus that could be uncovered through exploration, would become extraordinary profits for the companies, and not differential rents for the landowners.

Moreover, strictly speaking, landowners did not own the oil underlying their holdings. They only had the right to appropriate all of the petroleum that they could bring to the surface from wells on their lands – in the same way as they had the right to draw water from an aquifer or to hunt any wild animals that strayed on to their property, in accordance with what was known as the 'law of capture'. Therefore, the lessors obliged their lessees to drill wells on the boundaries of their properties, the idea being to both prevent oil under their land from being produced by their neighbors and to siphon off oil from under their neighbors' lands, a practice which also entailed producing maximum volumes as soon as possible. This was actually an important cause of the recurrent price collapses which tended to accompany any major discovery.

However, from 1929 on, the most important producing states, confronting one of the recurrent price crises, created official institutions to control production with a view to stabilizing prices in the name of *conservation*. Regulations were established covering the location of wells, and the minimal distance between them, regardless of the boundaries of surface property. This process went hand in hand with the transformation of the right of appropriation of the surface owners, as per the 'rule of capture', into a property right. They were to receive a royalty for their aliquot share in an oil pool – calculated on the basis of estimates by geologists, whose scientific discipline was developing in step with the petroleum industry – even if the petroleum in question were to be produced from wells located in neighboring lands.

Moreover, to improve recovery factors further, maximum efficiency rates for producing wells were established. These rates took into account the gas content of the oil produced as well as the structural characteristics of the reservoir. Maximum efficiency rates were to guarantee a stable flow of oil towards the wellhead, and to prevent its interruption due to excessively rapid production which could leave behind pockets of unrecoverable oil. In order to stabilize prices, the states would have the power to limit the output of more productive wells, while marginal wells would be allowed to produce at their full capacity, which was in any case quite modest, since limiting their output would have implied shutting them down, probably forever, given the high cost of bringing them back on stream again. In any event, when a well was shut down it had to be properly plugged. There is no doubt that conservation put an end to the most aberrant consequences of private mineral ownership.

In 1935, the conservation bodies set up by state governments united in what was finally called the *Interstate Oil and Gas Compact Commission*, with the support of the United States Federal Government. In fact, the Federal Government enacted laws that made the interstate trade of oil produced in violation of conservation measures a federal offense. For its part, the United States Supreme Court confirmed the constitutionality of these legislative measures, noting that they essentially involved the conservation of the natural resource and were not motivated by the monopolistic ambitions of oil producers. Thus, American domestic prices finally stabilized on the basis of the production costs of marginal wells and the corresponding proprietorial rent.

Furthermore, given the importance of both United States oil production and of United States companies to the international petroleum market at large, international petroleum prices were to be based on these domestic prices.

Federal regime

In essence, in all federal and state lands, as well as lands in Indian Reservations under the tutelage of the federal government, leases would follow the same patterns as in private landholdings. In any case, at the start of the 20th century, the states of the Union, as well the federal government, began to reserve for themselves the ownership of any hydrocarbons that might be found underground, even as surface landholdings continued to be privatized as part of the colonization of the country.

In sum, petroleum leases provided for surface rents and bonuses. The usual royalty rate would be one eighth (12.5%) in most of the country, but one sixth (16⅔%) in other parts. However, on public and private lands alike, the state governments, as overriding landlords, would usually levy a severance tax, economically equivalent to a royalty. In Texas, for example, the severance tax on the production of crude oil has always been 4.6%.

Clearly, given private property, rents and royalties constitute a ground rent. However, the situation is different in the case of lands in the public domain. If the oil and gas production from a given state in the Union were consumed locally, the population of that state would finally benefit from the corresponding rents and royalties, for instance, by the reduction of the general tax burden in this state. Hence, petroleum reservoirs in the public domain would definitely constitute free gifts of nature.

However, if production from leases on public lands were to exceed local consumption and therefore sold outside of the state, it would be out-of-state consumers who would ultimately foot the bill for the rents and royalties embedded in the price, as well as for any applicable severance tax – although unlike in-state consumers, out of state consumers would not obtain any benefit from the ensuing distribution of the proceeds by the exporting state. Thus, the population of the latter would effectively benefit from an interregional, interstate, or even international ground rent. Conversely, from the vantage point of consumers in states which were net importers, the fact that the natural resource in the state of origin was produced from private or from public lands was something that made no difference whatsoever.

With the increasing prominence of the petroleum industry in the 20th century, there arose American states and aboriginal reservations where petroleum, as an interregional, interstate or international source of ground rent, assumed great importance. This was the case, for example, with Texas and with the Osage reservation in Oklahoma. Nevertheless, even if the beneficiary state entities were invested with varying degrees of sovereignty, they were all under the hegemonic power of capital as wielded at a federal level from Washington D.C. There can be no question that, while accommodating the historical arrangement between capital and land, the petroleum policy of the USA as a whole was ultimately defined according to its domestic oil industry and its internationalization.

Income Tax and Depletion Allowance

The conception of mineral deposits as stores of a 'valuable commodity' acquired a special importance at the beginning of the 20th century with the introduction of income taxation. Accordingly, rents and royalties, at least in part, would represent the intrinsic value of the natural resource, and consequently in theory not subject to a tax applied to *net* income only.

However, the discoverers of these natural deposits of 'valuable commodities', in the case of petroleum, were the companies and, as such, it was not really far-fetched for them to claim a finder's reward. Indeed, the companies would insist that, in fact, they were the co-owners of the deposits discovered, together with the landowners and, in consequence, they were also entitled to benefit from the 'intrinsic' value of the mineral. This point of view came to prevail after the First World War, first in the case of petroleum and then spreading to all minerals. It achieved its definitive form in 1932, as the *percentage depletion allowance* (Lichtblau and Spriggs, 1952).

The companies took the position that this 'intrinsic' value was the root of *all* extraordinary profits that could be generated by the discovered deposits, which included, beyond rents and royalties, all differential rents, which Ricardo had explicitly set apart through his characterization of mineral resources, in their natural state, as 'valuable commodities'. The intrinsic value was estimated then to amount to 27.5% of the wellhead price, as opposed to the 12.5% or 16⅔% customary royalty rates that prevailed in different regions of the country.

Certainly, this 27.5% implied a very significant reduction in taxable income, all the more important when production costs were higher. In the 1950s, the income tax rate applicable to commercial activities in general was 52%, which was applied to the amount resulting from the subtraction of production costs from gross revenues (i.e. what is commonly referred to as *gross profit*). In contrast, in petroleum production, this rate was applied to the gross profit *minus* 27.5% of *gross revenues*.

Without a doubt, the driving force behind this legislation were the oil companies, and not the amorphous group of landholders, even though the latter would also benefit from the corresponding reduction in their taxable income.

Assuming a production cost close to zero – which would be the case for very successful leases but would also apply to the collection costs of rents and royalties of the landholders – the applicable rate would be decreased from 52% to slightly less than 37.7% (= 52%*(1 – 27.5%) = 52%*72.5%). But if we assume, for example, a production cost equal to 22.5% of the Price of the barrel at the wellhead, the applicable rate would be decreased to 26% (= 52%*(1 – 22.5% – 27.5%) = 52%*50%).

Furthermore, this 26% rate became legally the floor applicable to marginal discoveries given that, with higher costs, the 'intrinsic value' was reduced to 50% of the economic surplus – now always equal or lower than 27.5% of gross revenues – while the other 50% was considered as a profit for the producing companies.

Clearly, exploration and production were stimulated by being made more profitable, and it was this fact which justified politically the arrangement in question. On the other hand, the market price at the fields was supposed to be defined by competition at the national level. However, this was no longer really the case after 1959. First, domestic prices were maintained at higher levels as the government imposed import quotas in order to protect them from falling international prices. Then, starting in 1973, there was no option other than to let domestic prices increase with international prices. Indeed, the percentage depletion allowance was reduced, in 1969, to 22%, and would be further reduced to 15% by 1984. Furthermore, the right to enjoy the depletion allowance was subject to certain limitations meant to ensure that only small producers and royalty owners would qualify.

Peak production

Historically, the United States has been the most important petroleum producing country in the world. However, being an exhaustible natural resource, it peaked first in 1970, when United States production of crude oil and condensates reached 11.3 million b/d. Although the subsequent incorporation of Alaska into the ranks of the major oil producing states of the Union brought about a temporary arrest in the decline in production, it was only from 2009 onwards, as a result of the hydraulic fracking of shale formations, that the declining trend – albeit for a few years only – was reversed, moreover in spectacular fashion (Figure I.1).

Figure I.1 USA: PRODUCTION OF CRUDE OIL AND CONDENSATES 1922 – 2018

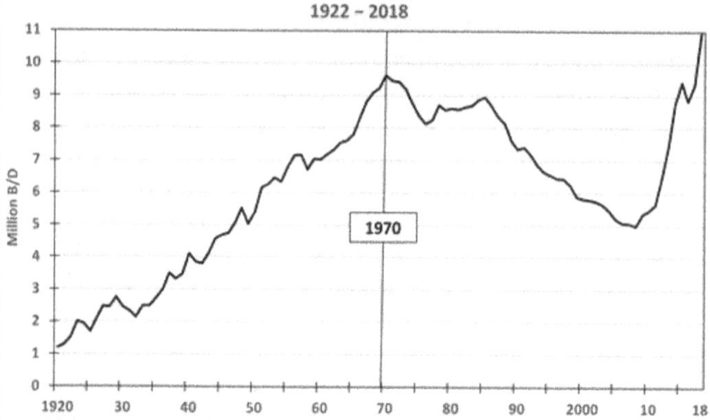

Source: American Petroleum Institute

3.2. Conclusions

The subordination of natural resources to capital manifested itself at the political, ideological, and theoretical levels by denying that there was any such thing as a proprietorial rent, and by postulating that any rent represented either a differential rent or, alternatively, the value of a 'natural' commodity.

Nevertheless, in the cases of British coal and American petroleum, private landed property – including minerals – became a serious obstacle to the development of productivity and, hence, of capitalism. In the case of US-American petroleum, this consequence was minimized by means of reforms. In the case of British coal, there were no reforms, which finally led to the expropriation of landowners. Expropriation was enacted in 1938, with an eye to the future development of productivity. Furthermore, from that point on, public debate would focus solely and exclusively on the productivity of the industry. Finally, when the coal industry was nationalized in 1946, the fact that the private ownership of the natural resource had been brought to an end in favor of public ownership, and that the latter was seen as definitively superior to the former, was completely suppressed. In due course, the industry would actually be re-privatized, but not so the natural resource.

Part II.

International Concessionary Regime

The development of capitalism in Great Britain led to a free trade policy predicated on exports of manufactured industrial products and imports of primary products, in such a way that international competition would put downward pressure on the prices of agricultural products. In this way, the scope for extraordinary surpluses, and therefore, for agricultural ground rent, would be narrowed down, first in Great Britain and ultimately throughout the whole of Europe (Kautsky, 1899).

This same policy also applied to extractive industries. In the case of petroleum, up until the First World War, the four most important producing countries in the world were the USA, Mexico, Russia and Romania. In all of them, the private ownership of minerals was the rule. However, the Mexican and Russian Revolutions put an end to this in 1917. Romania, for its part, put an end to private mineral ownership at the end of the Second World War, when it became part of the Soviet bloc. In Russia as well as in Romania, the nationalization of the natural resource went hand in hand with the nationalization of the industry; but in Mexico, the nationalization of the industry would only happen in 1938 (B. Mommer, 1999).

At present, private ownership of hydrocarbons only survives in the USA. In all the other countries, they form part of the public domain, and the international petroleum regime would have as its basis, until the beginning of the 1970s, a concessionary regime dominated by the European and US-American international oil companies.

The main importers were the developed capitalist countries, with the sole exception of the USA. While the USA would become a petroleum importer at the end of the First World War, a few years later it returned to being an exporting country, until 1947. But by a caprice of nature, the richest petroleum bearing lands were to be found in weak and dependent countries, and even though some of them were nominally independent, the majority were subject to semi-colonial or colonial conditions. In all of these countries, petroleum exports were to generate very significant extraordinary surpluses, not only in relative terms compared to the investments in petroleum themselves, but also compared to the size of the corresponding national economies.

There arose, then, a situation analogous to the one that has already been remarked upon in the case of the USA, in connection with both the interregional commerce of hydrocarbons by and among the federal states and the Indian communities. In the exporting countries petroleum was in the public domain and so, at a national level, it was a free gift of nature but, potentially, it could be a source of international ground rent. However, in the USA, the petroleum policy of the federal government followed the rules historically defined by the private sector, which meant that both Federal States and Indian communities were subject to these rules. In consequence, the scope they had to maximize interregional ground rent was, if not wholly non-existent, then very limited at best. However, the petroleum exporting countries were not subject to an international supreme authority comparable to that of the USA federal government.

Furthermore, in those countries where a superior colonial or semi-colonial authority existed, international petroleum politics would become part and parcel of the global struggle for independence. This struggle, furthermore, intensified markedly after the First World War.

The international petroleum policy of exporting countries with respect to the national ownership of the natural resource as a source of international ground rent ran into the limitations imposed by the developed capitalist countries and their international oil companies. The problem faced by the former was how to rid themselves of these limitations; whereas for the latter the problem was how to keep the exporting countries – in one way or the other – subordinate to the international development of capitalism and, ultimately, its globalization.

Petroleum as a source of international ground rent, throughout the concessionary period, surpassed in importance all other natural resources put together. In this process, the USA played the leading role, due to the extraordinary importance of its oil industry at the domestic level, to the prominence of its international oil companies, and finally, to its rise – after the First World War – to become the world's foremost power.

1. Middle East: International Arbitration

The concessionary regime in the Middle Eastern countries developed in the context of colonial and semi-colonial political structures, with the exception of the poor and backward Saudi Arabia, which became an independent country in the wake of the dissolution of the Ottoman Empire after the First World War. The legal basis of the concessions boiled down to bare title deeds, given the absence of any legislation covering the subject matter in the countries involved. In fact, the first mining law in the Middle East was enacted as late as 1957, in Iran. Moreover, the concessions were supposed to be subject to the principles of law common to the parties and the general legal principles of civilized nations. Any dispute would be subject to international arbitration, within an overarching legal structure created by the developed capitalist countries.

1.1. Iran (up to 1931)

The oldest among the major concessions in the Middle East had been granted by Iran (Persia, as it then was) to William K. D'Arcy, a British citizen. Iran was divided into two imperial areas of influence: Russian in the North and British in the South. The concession had a term of sixty years and it covered the whole of the southern part of the country, an area of 1.3 million km^2. Apart from a signature bonus of 20 thousand UK£ (about 97 thousand US$), and a package of shares (worth a similar amount) in the company that was to exploit the concession, the only relevant payment foreseen was a 16% profit share after the payment of UK income tax. As a matter of fact, this 16% was the sum of a 10% profit share plus a 6% additional charge which served as a proxy for general taxes. Moreover, Iran waived its sovereign right to tax.

The first significant discovery took place in 1908 and, in its wake, the Anglo-Persian Oil Company was established[1]. In 1914 – on the eve of the First World War – the British government acquired a majority shareholding in the company. Already at that point, the importance of petroleum for modern industries and, in particular, for the Royal Navy, was patently obvious.

1.2. Iraq (up to 1932)

At the beginning of 1914, British, Dutch and German interests agreed to request concessions within the Ottoman Empire jointly and not to compete among themselves, and pursuant to this agreement they organized themselves into the Turkish Petroleum Company. In mid-1914, this company obtained from the Turkish government the promise of a petroleum concession in Mesopotamia. After the First World War, with the collapse of the Ottoman Empire, the Turkish Petroleum Company claimed that this undertaking was still a binding one, albeit now on Iraq, which had passed under British mandate. Moreover, given the German defeat and French victory in the war, the German shares would come into French hands. However, as the USA had also been among the victorious powers – and it could be argued that it had emerged from the conflict as the foremost among such powers – and had just become an importing country, the American government demanded that its international oil companies also receive a stake in the Turkish Petroleum Company.

However, the European interests had already offered the Iraqi government a 20% shareholding in the Turkish Petroleum Company. In contrast, the US-American interests, true to the abhorrence felt in that country for state companies, put forward the idea of a royalty. Indeed, what the United States government had in mind was that, in Iraq, the example set by the federal government in the management of Osage lands would be followed. Ultimately, in 1925, the Turkish Petroleum Company obtained a concession extending over 91 thousand km^2, a fifth of Iraq's territory, for a period of 75 years. Even though the title deeds preserved a reference to a possible state shareholding of 20%, this was a dead letter. As far as the royalty was

---

[1] Renamed Anglo-Iranian Oil Company in 1935, and British Petroleum in 1954.

concerned, a figure of four shillings (gold) per ton was agreed, equivalent to around 0.13 US$/b. This amount, at the price levels prevailing at the time, was equivalent to a 1/8 royalty. It was to be adjusted twenty years after production had started, depending on the profitability of the Turkish Petroleum Company, but only up to a maximum of six, or down to a minimum of two, shillings (gold) per ton.

1.2.1. General Taxation in the Non-Oil Sector

The Turkish Petroleum Company would also be subject to general taxation, albeit only in the same measure as the non-petroleum sector (Stocking, 1970, p. 132). This arrangement was not to the liking of the companies, and the existence of such a clause was due to the tortuous negotiating process around this concession. Besides the participation of the international petroleum companies which competed with one another, the governments of Great Britain, France and the USA had also become involved. In fact, a definitive agreement with regard to the various shareholdings in the Turkish Petroleum Company – now renamed the Iraq Petroleum Company – was only reached in 1928. Each of British Petroleum, Royal Dutch-Shell, Compagnie Française des Pétroles and an US-American consortium constituted by the Standard Oil Company of New Jersey and the Standard Oil Company of New York, would hold a 23.75% interest[2]; the remaining 5% of the shares would be held by the intermediary who had brokered the deal.

Once the Iraq Petroleum Company had achieved firm control over the concession, it insisted on renegotiating the general taxation clause. Finally, in 1931, the government of Iraq accepted that the general taxes for the non-petroleum sector be estimated, once and for all, for the next 69 years, at four pence per ton (0.33 shillings). At the same time, a secure rent of 400 thousand UK£ (gold) per annum – 1.9 million US$ – was agreed, to be paid before production reached significant levels. The following year, Iraq was granted its Independence by Great Britain, and exports started in 1935.

1.3. Iran (1930-1933)

In the meantime, Iran, with one sole immense concession in the hands of a British state-owned company, was still the sole petroleum producer in the Middle East. The fiscal participation that had been agreed upon at the turn of the century was modest in comparison with the recent arrangements in Iraq. Nevertheless, when in 1930 the Iranian government introduced a 4% income

[2] At the time, both British Petroleum and Compagnie Française des Pétroles were State-owned.

tax, the Anglo-Iranian Oil Company refused to pay insisting on the fact that Iran had waived its tax sovereignty granting the concession.

In 1931, British Petroleum produced 5.73 million tons of petroleum (approximately 115 thousand b/d), but the fiscal take of the Iranian government – in the midst of a world petroleum market in acute recession – came to barely 310 thousand UK£ (1.5 million US$). That same year, the Iraq Petroleum Company paid 400 thousand UK£ (1.9 million US$) on account of the recently agreed secure rent, without a single barrel having been produced yet. The Iranian government now proceeded to rescind the D'Arcy concession. The response was that British gunboats were put on alert to patrol the Persian Gulf, to prevent the export of 'stolen' oil. But Great Britain also appealed to the Society of Nations in Geneva, where both parties presented their respective cases (Société des Nations, 1932/3). Ultimately, British Petroleum accepted the – long standing, in fact – Iranian proposition that the concession be renegotiated.

From a legal point of view, the outcome of the renegotiation would be a new concession, which would have no connection with the old one. As part of this wiping the slate clean, the Anglo-Iranian Oil Company paid one million pounds sterling in full and final settlement of all past claims. The new concession was granted in 1933, with retroactive effect to 1931, for a further term of 60 years. The area under concession was reduced to 260 thousand km^2 and, for its part, the government renounced to have any shareholding.

Following the patterns established in Iraq, a secure rent of 750 thousand US$ was agreed. The Anglo-Iranian Oil Company would be subject to 'general taxes' levied on the non-oil sector, which were contractually fixed for thirty years and were to be renegotiated in 1963. Most importantly, a royalty of four shillings (gold) per ton was agreed. However, when the title deeds of the new concession were signed, in 1933, the British pound had just been devalued by 29%. Thus, in nominal terms, Iraq would now receive a royalty of five shillings and eight pence, whereas Iran would only get four shillings.

There is no doubt that, given that the concession was already producing and had demonstrated an extraordinary profitability, the Iranian government obtained terms far short of those that would have been warranted, and British imperial might had prevailed.

1.4. Bahrein, Kuwait, Qatar and Saudi Arabia (1930-1935)

In 1928, as soon as the shareholding structure of the Iraq Petroleum Company had been finalized, the three largest oil companies in the world at the time – Royal Dutch-Shell, British Petroleum and SONJ (Standard Oil of New Jersey, later Exxon) – set up the International Petroleum Cartel. They agreed not to compete among themselves in obtaining concessions within what had been the boundaries of the former Ottoman empire, in the same way as had been agreed before in 1914 by the original participants in the Turkish Petroleum Company. Subsequently Gulf, Texaco, the Standard Oil Company of California (Chevron) and the Standard Oil Company of New York (Mobil) joined this cartel of concessionaires (United States Senate, 1952).

Between 1930 and 1935, these companies obtained a further three major concessions in the Middle East which resulted in significant strikes before the outbreak of the Second World War; to wit, in Bahrein, Kuwait and Qatar. These small sheikhdoms were British protectorates at the time and they granted the whole of their respective national territories. The agreed royalty was four shillings and six pence per ton, but payable in Indian currency and without any gold backing (and hence subject to the devaluation of this currency).

In the meantime, at the start of 1933 (and just before joining the International Petroleum Cartel), the Standard Oil Company of California had managed to obtain its famous concession in Saudi Arabia, for a term of 60 years and covering an extension of 1.3 million km^2. Saudi Arabia was the only nominally independent country to have emerged from the decomposition of the Ottoman empire, and it managed to obtain a royalty of four shillings (gold) per ton, at the gold/sterling parity that prevailed before the subsequent devaluation of the British currency.

In each one of these concessions, by the by, the respective state waived its tax sovereignty.

2. Venezuela: Exclusive National Jurisdiction

Since 1784 mining in Venezuela had been subject to the Mining Ordinances of New Spain (Ordenanzas de Minería de Nueva España), which reserved the minerals to the Crown, "bitumens or juices of the earth" included (González Miranda, 1958, p. 33). The existence of hydrocarbons in Venezuelan territory was well known: in the west in the vicinity of Lake Maracaibo, and in the east on account of the Guanoco Asphalt Lake. In 1829, with Independence, the Liberator Simon Bolívar validated the Ordinances by decree, substituting the figure of the 'Crown' for that of the 'Republic'. Subsequently, the Republic was to develop its own mining codes and statutes which, towards the end of the 19th century, followed the patterns of the most modern legislation at the time: i.e. French mining laws (García Macías and I. Mommer, 2018).

The first four petroleum concessions of future importance were granted in 1907, based on the Mining Law (Ley de Minas) of 1905, which provided for exclusive national jurisdiction: "[…] for no reason nor for any cause shall they give rise to diplomatic actions or foreign claims" (art. 6). Moreover, foreign state-owned companies were barred from acquiring concessions.

The concessions were subject to the state sovereign power of taxation, although, of course, the specific mining taxes defined by the law and in title deeds took on the character of vested rights. In any event, the Venezuelan executive was entitled, at its sole discretion, to exempt the concessionaires from the payment of import duties, provided that the materials imported were directly related to the production and export of petroleum from the concession.

The concessions ran for terms that could be as long as fifty years, and the areas granted were very extensive. However, the exploration term was limited to just a few years, and once it expired, concessionaires had to choose production parcels with a maximum extension of 200 hectares, subject to a surface tax of 2.00 bolivars per hectare (Bs/ha) and an exploitation tax of 4.00 bolivars per ton (Bs/t). The gold parity of the bolívar was then 5.20 per US$. Thus, the 4.00 Bs/t were equivalent to, approximately, 0.11 US$/b. The surface tax was meant to ensure that at the end of the exploration term, the concessionaires would return all those areas in which they did not have a serious interest.

2.1. Waiving Tax Sovereignty

These first concessions were granted by the government of General Cipriano Castro (1899-1908). In December 1908, Castro was ousted by General Juan Vicente Gómez, with the explicit backing of the USA. In 1909, Gómez promulgated a new Mining Code (Código de Minas) which set forth that "any mining concession title is a contract between the National Government and the concessionaire with regard to the rights and obligations provided for in the present Law, taxes inclusive" (art. 23; emphasize ours). In the Exposition of Motives, it was stated that concessionaires only had to pay taxes, like every other economic agent, and that the applicable mining taxes were nothing other than a specific type of general taxation. The concessionaires, then, were not expected to provide any compensation for the natural resource itself, considered to be a free gift of nature, completely in accordance with French mining legislation.

Mining taxes were halved and concessionaires were definitively exempted from having to pay any import duties. In the subsequent Mining Code (Código de Minas) of 1910, which was of a very similar ilk, the incumbent concessionaires were offered the option of bringing their operations under its (more favorable) terms, and all of them chose to do so. All the important concessions whose legal basis was the 1910 Code would eventually end up in the hands of Royal Dutch-Shell.

2.2. From the French to the American Reference

Venezuela became a petroleum exporting country in 1917. That was the same year that United States oil companies first turned up in the country – the USA had just become an oil importing country – where they would challenge the established companies with offers that reflected their experiences as lessees in their country of origin, ferociously competing in the process with the incumbent European companies. Thus, the Venezuelan government also became aware of the fact that, as well as being a magnet for foreign investment, the country's petroleum endowment could also become a source of international ground rent. As a result, the government abandoned the French reference and instead adopted the US-American one, which would be used to develop a specific hydrocarbons legislation.

The first Hydrocarbons Law (Ley de Hidrocarburos), from 1920, followed the patterns of United States legislation applying to Federal lands. It was followed by the 1921 and 1922 Hydrocarbons Laws. The latter once again gave incumbent concessionaires the chance to bring their concession titles in line with it. All concessionaires with title deeds dating from 1918 onwards took this option, because the new law reduced rents and royalties by half in comparison to the 1920 Hydrocarbons Law. Subsequently enacted Hydrocarbons Laws, until 1935 (when General Gómez died), differed only in minor details from that of 1922, so we can here concentrate on summing up the latter.

The exploration-exploitation concessions had a term of forty years and covered a maximal extension of ten thousand hectares, divided into parcels of 200 hectares each. At the end of the exploration term, of up to three years, the area of the concession had to be reduced in half, and a surface rent (or, in strict legal terms, a surface tax) of one to two Bs/ha was payable. Royalty (in strict legal terms, the exploitation tax) was set at 10% for onshore tracts but only 7.5% in submerged tracts. However, and quite prudently, both onshore and offshore tracts also had guaranteed minimum royalties of 2.00 Bs/t and 1.50 Bs/t, respectively. The returned areas were to constitute national reserves which would be put up for auction later on, as exploitation concessions, with a view towards obtaining higher rents and royalties. Indeed, royalty rates as high as 15% were obtained in such auctions. Furthermore, in all Hydrocarbons Laws up to that of 1935, the state waived its sovereign right to tax.

In essence, the concessions granted from 1918 onwards ended up in the hands of Standard Oil of New Jersey and Gulf Oil. In 1928, Venezuela became the foremost petroleum exporter in the world, a position that it would hold for 42 years, until 1970.

2.3. Petroleum Reform of 1943

When General Gómez took office in 1908, Venezuela was one of the poorest countries in Latin America; by the time he died, in 1935, Venezuela was already a prosperous oil country. Initially, the country had benefited mainly from a huge capital influx and then, as an exporter of petroleum, it would continue benefiting from a significant international petroleum rent. Furthermore, the country was also reaping the benefits of a significant immigration of human capital. Political and social structures in the country developed accordingly and, in 1936, the first modern political parties were set up.

Consequently, the general exoneration from import duties enjoyed by the concessionaires began to draw increasingly loud opposition. But when the government attempted to change this situation, the companies turned to the Supreme Court to defend what they saw as their vested rights, and they did so successfully. However, the international oil companies were definitely subject to the tax sovereignty of their respective countries of origin – the USA, Great Britain, and the Netherlands – and, in particular, to each country's corporate income tax, which was also applied to their Venezuelan profits. The Venezuela government considered this situation to be unacceptable. Referring to a new national petroleum policy, the President of the Republic, General Eleazar López Contreras, justified the reform to the National Congress in the following words:

> [...] the action of the Public Power has to be directed towards achieving the right that the state has to the greatest possible participation in the riches of its subsoil [...] (Egaña, 1939, p. XI; emphasize in the original).

The government reacted to its defeat in the Court by suspending the granting of new concessions and the Minister of Development, Manuel R. Egaña, commissioned a study on the legal defects that could be discerned in some of the major concessions. Between 1913 and 1940, a total of 4.208 exploration-exploitation concessions had been granted, as well as 4.471 exploitation concessions covering areas of the national reserves.

In July 1942, President (and General) Isaías Medina Angarita announced that his government would undertake a root and branch petroleum reform. Although SONJ initially took a confrontational stand against the reform, the company fell in line on orders from Washington. The US-American government was not willing to run the risk of a debacle in Venezuela like the one that had happened in Mexico only four years ago, at the height of the Second World War (in that same year German submarines sunk a number of tankers carrying Venezuelan oil, and shelled the Curaçao refinery). As a matter of fact, the United States government sent two of its top experts in the subject matter to monitor the reform.

2.3.1. Hydrocarbons Law

The 1943 Petroleum Reform involved, first of all, a new Hydrocarbons Law, which reflected the best practices in federal lands in the USA, with rents and royalties at the corresponding customary levels. Royalty would be one sixth and, prudently, minimum prices for its liquidation were also set; concretely, at the average prices prevailing before the outbreak of the Second World War, from 1937 to 1939. Surface rents were set at 5.00 Bs/ha which, with an exchange rate at gold parity of 3.09 Bs/US$, came to 1.62 US$/ha[3].

The new law offered the incumbent concessionaires the option of bringing their concessions under its purview, in exchange for which they would obtain new concession titles for a further forty years, wiping the slate clean of any faults and defects that they might have had. Practically all the concessionaires availed themselves of this offer. In so doing, they accepted an immediate increase in the applicable mining taxes: The exploitation tax had averaged only 9.3% in 1942 but would now be 16⅔%. Moreover, the areas under concession were reduced from 7.1 million ha to 5.5 million ha.

Royalty Liquidation Prices

In line with established United States practice, the law provided that the state as grantor of the concession had the right to receive the royalty in kind and to sell it for its own account. The concessionaires, for their part, would be obliged to render the services necessary to deliver the oil in question to the designated buyer. With this arrangement, the state could be assured that the royalty liquidation price would be the highest prevailing market price. As a practical matter, the government resorted to this option on various occasions during the first years of the post-war period. Nevertheless, from 1947 onwards, when the USA became an importing country, the best prevailing prices would be the posted prices of the United States market. From that point on, the government signed royalty agreements with the concessionaires on the basis of such prices. That meant, from a legal point of view, that the concessionaries would purchase the royalty volumes at the agreed prices.

2.3.2. Industrialization of Petroleum

Between 1918 and 1932, Shell and SONJ had built large refineries in the Dutch islands of Curaçao and Aruba – located just off the Venezuelan coast – in order to process their Venezuelan production. For the promoters of the Petroleum Reform, this was an unacceptable situation, and they demanded not only that the two companies put an end to the expansion and modernization of their refining capacity in these islands, but also that they build large refineries in Venezuela, within a five-year period counted from the end of the war. This proved to be a highly contentious point between the parties, and was the very last one on which an agreement was reached, and then only because the government threatened to break off negotiations and take unilateral measures.

---

[3] In 1934, the USA devalued the dollar but Venezuela did not devalue the bolivar. Thus, the exchange rate at gold parity went from 5.20 Bs/US$ to 3.09 Bs/US$

2.3.3. *Tax Sovereignty*

The 1943 Hydrocarbons Law reasserted that mining taxes would remain in force throughout the whole of the duration of the concession. Just as in the Mining Law of 1905, however, the Executive was given the faculty to exempt the concessionaires, at its sole discretion, from the payment of import duties, provided that the imported goods were directly related to the purpose and object of the concession.

But the Law stipulated that "concessionaires shall pay all general taxes, of whatever kind, and shall also pay the legal charges, contributions, and dues for the services that may be rendered them" (art. 46)[4]. Hence, the concessions were subject, in particular, to the new Income Tax Law, which came into force in January 1943. For the time being, the applicable income tax rate for the petroleum sector was 12%, equivalent to the effective income tax rate in the USA before the outbreak of the Second World War. Given United States legislation on double taxation, the result would be that the United States tax obligations of the US-American concessionaires would be reduced by the amount of income tax that they would pay in Venezuela. Both Great Britain and the Netherlands would finally follow the same guidelines.

2.3.4. *New Concession Bidding Rounds*

SONJ and Royal Dutch-Shell took advantage of the renovation of concessions in order to re-organize their Venezuelan affairs. The former consolidated its Venezuelan interests into the Creole Oil Company, and the latter into the Shell de Venezuela Company, while Gulf interests had already been organized into the Mene Grande Oil Company. In 1943 these three large companies controlled 58.7%, 28.8% and 8.7%, respectively, of Venezuelan oil production (for a collective share of 95.2%). Furthermore, in 1936 and 1937, SONJ and Shell each acquired a 25% share in Mene Grande, a company that effectively functioned as the coordination center for their production in Venezuela. This arrangement was part of the structure underpinning the International Petroleum Cartel.

With the reform process concluded, the Medina Angarita government went ahead with granting new concessions. However, it viewed the cartelization on the part of the concessionaires as being inimical to the national interest, since competition among the firms, as aspirants to obtain exploration-exploitation concessions, was an essential requisite if the government was to obtain a ground rent that reflected market conditions. For this reason, the 1943

---

[4] All quotes related to the 1943 Hydrocarbons Law are taken from its translation by Gilbert Grace Cover, published by the Embassy of Venezuela Information Service, Washington D.C.

Hydrocarbons Law declared petroleum pipelines to be public services (common carriers), a move meant to guarantee access to the market of new competing concessionaires.

In the bidding round of 1944/5 royalty rates as high as one-third were obtained, and bonuses came to 58 million US$. The extension of the new areas granted in concession was 6.3 million ha, meaning that the total area under concession by 1945 was 11.7 million ha. By 1955, this figure had gone down, on account of both the Law and the economic pressure of the surface tax, to 5.9 million hectares. The share of the three largest companies in Venezuelan production was now 88.4%.

One of the conditions for participation in the bidding rounds was that the output of the new concessions would have to be refined either in Venezuela or else somewhere beyond Aruba and Curaçao. However, the ambassadors of Great Britain, the Netherlands and the USA objected to this provision, arguing that it was an affront to liberty of commerce. President Medina reacted to these protests by ordering the immediate suspension of the bidding rounds then in progress, which were only resumed once the ambassadors had withdrawn their protests.

The last bidding round for concessions was to take place between 1956 and 1957. A total of 821 thousand hectares were awarded to the highest bidder; bonus payments reached the sum of 712 million US$ and royalty rates as high as 25.5% were achieved. At the same time, the collaboration of some of the participants in the bidding rounds was obtained in order to develop a domestic petrochemical industry. The share of the three major concessionaires in Venezuelan production continued to shrink, and by 1970 was down to 80%.

2.3.5. Conclusions

With the 1943 Reform, the Venezuelan petroleum regime caught up with the one prevailing in the United States. However, the Venezuelan concessions had a finite term, which was not the case with petroleum leases in the USA. Moreover, the USA petroleum regime was subject to the hegemonic power of capital, albeit against the backdrop of private mineral ownership; whereas in Venezuela the state came from imposing itself as Nation State, as the sole and sovereign landlord.

However, neither the United States government nor the oil multinationals were willing to accept that the Venezuelan state could go beyond the rents and royalties defined in the 1943 Hydrocarbons Law and the title deeds, and avail itself of its tax sovereignty to increase its share in extraordinary surpluses. Indeed, in a memorandum commenting on the draft bill for the 1943 Hydrocarbons Law and addressed to the Venezuelan government, the US-American advisors asserted categorically that:

> [...] a royalty of one sixth for the owner of the reservoirs represents very nearly 50 per cent of the net benefits derived from the exploitation of such fields. It cannot be expected that any owner might demand more and, certainly, no company would be willing to give more than this equitable participation (Curtice and Hoover, 02/01/1943).

They therefore insisted that in the Hydrocarbons Law, the phrase already quoted – "concessionaires shall pay all general taxes, of whatever kind, and shall also pay the legal charges, contributions, and dues for the services that may be rendered them" – be qualified by the following colophon: "which have the same impact on all enterprises". But the Venezuelan government rejected this proposal, and the 1943 Hydrocarbons Law definitively asserted its sovereign right to levy taxes, beyond mineral taxes.

Nevertheless, the Exposition of Motives of the 1943 Hydrocarbons Law did include a reference to a fiscal participation of 50%:

> According to the numerous and painstaking calculations carried out, that 16⅔% applied to exploitation, together with all other taxes, is the equivalent, in our country, approximately, on average, to 50% of the benefits of the extractive petroleum industry (Emphasize ours).

This reference to "other taxes" was somewhat ambiguous. Was it making an allusion to other mineral taxes, or to other general taxes? As far as the United States advisors were concerned, they stated that:

> [...] the application of the new Income Tax Law has not been part of the calculations which determine the production costs of the oil industry. This sort of tax, as applied in other countries, is used as a control over excess profits obtained by individuals and commercial enterprises. If used wisely this tax can become another guarantee for the Nation regarding the concept of its fair share in the benefits of the Venezuelan industry (*El País*, 28/12/1946; emphasize ours).

With the inclusion of income tax, it was estimated that the share of benefits would be close to 60:40 in favor of the Nation.

2.4. The Fifty-Fifty Arrangement

In 1943, according to the Income Tax Law, the rate applicable to the petroleum sector was 12%. This rate was equivalent to the effective Federal income tax rate for hydrocarbons producers in the USA, before the Second World War. However, after the end of the War, the USA maintained income tax rates at a high level. Inevitably, Venezuela was going to consider increasing its own income tax rate accordingly.

In October 1945, a civilian-military coup brought down the government of General Medina Angarita. The military wing of the conspiracy was headed by

Major Marcos Pérez Jiménez and the civilian component was constituted by the Acción Democrática (AD) party, under the leadership of Rómulo Betancourt, who had founded the party in 1941. He would now preside the Government Junta. As Minister for Development – and, as such, responsible for petroleum policy –, he named Juan Pablo Pérez Alfonzo, a lawyer.

On December 31, 1945, the Government Junta decreed an extraordinary tax on high incomes, which was to be applied that year only. The following year, after the National Assembly had been elected, the Junta promoted a corresponding reform of the Income Tax Law. The rate applicable to the highest income of the oil companies was increased to 28.5%. This rate was essentially equivalent to the effective rate the United States oil companies would have to pay, in any event, in the USA.

In its message explaining the initiative to the National Assembly, the government stressed that this fiscal reform had as its foundation the 1943 Petroleum Reform and the corresponding discussions about tax sovereignty and a 60:40 participation of the state in the benefits (B. Mommer, 1988, p. 85f). Furthermore, before the National Congress, the government insisted that:

> It was the decision [...] of the government [...] not to reach the sought-for goal by means of the enactment of an extraordinary tax. Instead, [we propose] a reform in the direct tribute regime which could be analyzed and debated by the National Assembly [...] depositary of the National Sovereignty; and not to resort to a formula of transitory validity, but to a modification in the scaling of the tax [...] which could be in force for a number of years. In our conception, and without implying thereby any lack of understanding that the faculty to tax constitutes one of the essential attributes of National Sovereignty, the Reform which we present for the consideration of the National Constituent Assembly is capable of ensuring for the long term an equitable participation for the state and the Nation in the profits obtained by extractive industries (*El País*, 28/12/1946).

However, due to the continuous increase in both output and prices, the reform to the Income Tax Law fell somewhat short in hindsight. In 1946 and 1947, the state's share of benefits as regards Creole and Shell – the companies with the best lands – did not even reach a ratio of 50:50. Thus, in 1948, the government (actually, the government-elect, with Rómulo Gallegos of AD as President and Pérez Alfonzo continuing as Minister of Development) went back to the National Congress and proposed yet another reform to the Income Tax Law, which would create an Additional Tax that would apply to excess profits, and which would take the fiscal participation of the state to 50% of extraordinary petroleum surpluses. Furthermore, in the Exposition of Motives the government alleged that:

> The application of the 1946 reform to the profits recorded that year demonstrated just how impractical it had been to amend a tax scale which should have remained in accordance with the desired aim of obtaining for the Nation a 50 per cent participation in the profits (Emphasize ours).

As a matter of fact, in 1946, the government had made no reference to a ratio of 50:50, but to 60:40, income tax included. But now the government referred to a profit share limited to 50% levied only on profits in excess of 15% of invested capital, after income tax. Moreover, the government informed the National Assembly that Creole and Shell had volunteered to liquidate that Additional Tax, with retroactive effect, for the years 1946 and 1947. Indeed, in 1952, Creole made a retroactive payment of 4.8 million US$, while Shell would pay 1.4 million US$ in 1953 (Vallenilla, 1973, p. 200). These were sums of little importance in comparison to the 382.8 million US$ of petroleum fiscal income recorded for those two years.

2.4.1. The Historiography of Fifty-Fifty

How these voluntary retroactive payments were to be understood, from the standpoint of the major concessionaires, comes across in the article "Creole Petroleum: Business Embassy", published by Fortune magazine in February 1949:

> In 1941, when a new General-President, Isaías Medina, took office, government officials demanded a […] greater participation in the industry. The new government announced the principle that royalties and other general taxes should give the nation an oil income equal to the net income of the companies. […]
>
> Creole says today that it did not accept the fifty-fifty principle at the time, but it did agree to the higher royalty rate […].
>
> The loudest opposition to the legislation came from a newly legalized minority party known as Acción Democrática […]. The A.D. delegation in Congress asked for no more than the fifty-fifty share but argued that the 16⅔ per cent royalty formula would not give the government an even break if the price of oil […] should rise materially (*Fortune*, 02/1949; emphasize ours).

In fact, it had been the US-American advisors, and not the government, who had made a reference to a 50:50 profit sharing, and this ratio furthermore excluded general taxes and, in particular, income tax. For his part, in 1943, the main spokesman for Acción Democrática in the National Assembly and member of the Development Commission, Juan Pablo Pérez Alfonzo, in his Dissenting Vote never even mentioned a 50:50 split. He only made a critical reference to the estimates of the government, according to which mineral taxes would result, for a start, in a 60:40 split in favor of the country. Neither did he so much as mention income tax. What he did say was that all petroleum concessions in general, and not only those related to the transportation by pipeline, constituted public services, which in turn meant that concessionaires were only entitled to a normal profit, whereas all extraordinary surplus had to be transformed into fiscal income (Pérez Alfonzo, 1943).

Furthermore, under the subtitle Mr. Proudfit's Fifty – Proudfit was the President of Creole at the time – the article in question made reference to the Additional Tax:

> Creole's own tax experts, joined by others in the industry, themselves cooperated in drafting legislation requiring them to split the profit difference with the government in any year in which normal taxes and royalties leave the companies with a bigger share than the tax collector.
>
> Creole and Shell [...] express themselves as reconciled with this deal. [...] In general [...] the oil industry – and Creole most emphatically – does not worry so much about the effect of the fifty-fifty tax law as it does about future politicians who may have sixty-forty or even seventy-thirty ideas (*Fortune*, 02/1949).

In other words, the conceit was that, in 1948, a democratic government had managed to obtain what an oligarchic government had failed to obtain previously during its time in office. In the international arena, the oil companies would take on the job of imposing this vision; in the national arena, this vision was defended by the AD leadership, and in particular Rómulo Betancourt and Juan Pablo Pérez Alfonzo. Hence, the 1943 Reform was stripped of all significance and, instead, the fifty-fifty arrangement of 1948 was mythologized. But the fact of the matter is that, between 1946 and 1956, the Additional Tax receipts came to barely 112 million dollars, whereas the ordinary petroleum fiscal income, excluding bonuses, amounted to 4.4 billion dollars.

2.5. The Venezuelan Delegation to the Middle East

In 1949, the world petroleum market witnessed a brief, albeit significant recession, and production in the Middle East for the first time overtook that of Venezuela. The fields in the Middle East were much more bountiful, and the rents and royalties were much lower. In the USA average production rates per well were in the tens of barrels per day, while in Venezuela they were counted in the hundreds. In the Middle East they were counted in the thousands. There was therefore no question that oil from the Middle East could pose a serious threat to Venezuela. In March 1949, Joseph E. Pogue, at the time vice-president of the Chase National Bank in New York, delivered in Caracas a well-publicized lecture in which he warned that the pressure of competition from the Middle East might force Venezuela to reduce both its labor costs and its taxation levels, which at the time were among the highest in the world (Pogue, 06/1949).

Among the audience was Venezuela's Minister of Development, once again Manuel R. Egaña. Profoundly worried, he decided to appoint – together with the Foreign Minister – a diplomatic delegation that should visit Saudi Arabia, Egypt, Iran, Iraq, and Kuwait. The idea behind this initiative was to look for a way to "achieve an equilibrium of competitive forces, by means of which it would be possible to obtain benefits for the peoples of the Middle East but without the economic position of the Venezuelan people suffering any detriment" (Egaña, 1949). The delegation would provide copies and translations of all the relevant Venezuelan legal texts and would also extend invitations so that delegates from the countries to be visited could participate, as observers, in the First National Petroleum Convention, which had been scheduled by the government and the oil industry to take place in the near future.

3. The Fifty-Fifty Arrangement in the Middle East

3.1. Saudi Arabia (1945-1950)

As soon as the Second World War was over, Standard Oil of California set up the Arabian American Oil Co. (Aramco), as a joint enterprise in which it held a 30 percent share and that also included Standard Oil of New Jersey (30%), Texaco (30%) and Standard Oil of New York (10%). In the Middle East, Saudi Arabia was the only petroleum exporting country where only United States oil companies operated.

With a fixed royalty per ton, the Saudi government was going to derive no benefit from rising prices. But already in 1948, Aramco agreed to increase the royalty from 0.21 US$/b to 0.32 US$/b. This pre-empted the anticipated increase in the per barrel royalty in Iraq, from four to six shillings. Despite this, Aramco would pay a total of 32 million US$ to the Saudi treasury, whereas it paid 43 million US$ in income tax to the American government. Then a new arrangement was agreed, to enter into force from 1950 onwards. On the one hand, the fixed royalty of four shillings per ton was converted into an ad valorem royalty of one eighth. On the other hand, the Saudi government enacted an Income Tax Law, with a basic rate of 20% as well as an Additional Tax of 50%. These modifications were duly incorporated into the concession contract. In 1950, the government take thereby rose to 0.68 US$/b, while at the same time Aramco ceased to pay any income tax on its profits from Saudi Arabia in the USA.

The Venezuelan delegation played no part in this development of the fifty-fifty, as the Saudi government had refused to grant visas to the Venezuelan representatives.

3.2. Iran (1947-1954)

In Iran, the royalty was significantly smaller than in Saudi Arabia. Furthermore, the Iranian government was aware of the evolution of the petroleum regime in Venezuela, which was reflected in the proposal advanced in 1947, to bring the fiscal regime applicable to the Anglo-Iranian Oil Company in line with the Venezuelan fiscal regime.

In October 1949, the Venezuelan delegation reached Tehran. They were received by Manucher Farman Farmaian, a high official in the Ministry responsible for petroleum. The delegation was under instructions not to put forward any questions, in order to avoid raising suspicions, but they were told they should answer any questions put to them. The Iranian officials displayed an insatiable curiosity, and their questions covered all relevant subjects regarding petroleum (Farman Farmaian, 1997).

But the Anglo-Iranian Oil Company, confident in the colonial power of Great Britain, imposed a proposal that implied a 30:70 split in its favor. This agreement still had to be ratified by the Iranian parliament, which refused to do so in December 1950. Then, at the beginning of 1951, when it became public knowledge that Aramco had reached a fifty-fifty arrangement with Saudi Arabia, the government of Muhammad Mossadegh nationalized the Anglo-Iranian Oil Company, and the National Iranian Oil Company took over the former concession.

This gave rise to an international embargo against Iran, with the participation of the United States, France and the Netherlands as well as Great Britain, which no longer had the strength to enforce such a policy on her own. Finally, the Mossadegh government was overthrown in 1953 in a coup promoted by the USA and Great Britain. Formally, though, the nationalization was not reversed; instead, an International Consortium was formed to be the operator of the former British Petroleum concession, working as a contractor to the assignee of all exploration and exploitation rights, the National Iranian Oil Company. What did come to an end was the exclusive British colonial domination over the country, and British Petroleum only retained a 40% shareholding in the International Consortium. US-American companies would hold another 40%, while Royal Dutch-Shell and Compagnie Française des Pétroles were assigned 14% and 6% of the shares, respectively. The term of the 'operator' was to last for 25 years, with an extension option for a further 15 years (which would take the arrangement to 1993). The International Consortium would be subject to an equivalent of a fifty-fifty, in line with the pattern pioneered by Aramco in Saudi Arabia.

For its part, the British government became aware of just how important the intermediation of a national oil company could be in the process of negotiations with the government:

[…] there might even be an advantage in the international oil companies discreetly encouraging the formation of national oil companies […]. There seems little doubt that understanding by the oil-bearing countries of the technical and practical problems of oil operations is on balance more advantageous to the operating companies than ignorance and myth. By its nature and organization, a national oil company is more likely to gain the necessary knowledge, experience and authority to deal with oil practically and sensibly than a purely governmental administration, which would be more subject to direct political pressures and inhibitions (in Burdett, 2004, p. 280; emphasize in the original).

3.3. The Fifty-Fifty Mythology

Among the countries of the Middle East, Saudi Arabia was the first to benefit from a fifty-fifty, and Iran was the last. In its conception, the fifty-fifty was quintessentially US-American. All the same it was necessary to dress it up both politically and ideologically, so that it could be presented as an inherently equitable and fair arrangement. In the words of the SONJ chairman and chief executive officer:

> Fifty-fifty relieves the country of any financial risk, and places that risk upon the company which is in a position to evaluate it and spread the risks of the search for oil over many areas. Any basis other than 50:50 could create an imbalance of interests which would reduce the attractiveness of the venture for one or the other of the parties. The 50:50 represents a tested principle for maintaining an equality of interests through all the aspects of an inevitably complex relationship intended to endure for many years (Rathbone, 06/03/1958).

Furthermore, according to H. W. Page (a SONJ executive director), the roots of the fifty-fifty in the Middle East were supposed to reach as far back as the first concession granted in Iraq:

> […] in the Middle East pre-war […] 4 s. gold […] plus other payments to the governments, came to an average of nearly 50% of the profit.

However:

> After the War, with […] the rapid increase of crude prices, higher United States and United Kingdom corporate income taxes […], the oil companies operating in the Middle East recognized that the fixed royalty payment no longer gave the equitable division originally intended, and that additional royalty payments were neither economically practical nor a permanent method of maintaining equity between the parties. The […] 50:50 was the result of this recognition of the need to restore the equity which had been frustrated by drastically changed conditions, beyond the control of either party and unforeseen at the time the agreements were negotiated (Page, 1960).

Neither in the negotiations about the first concession in Iraq nor in the negotiations about the Saudi concession was there ever even a passing mention to a fifty-fifty. But the idea behind this re-writing of history was to take the fifty-fifty as far back into the remote past as possible, while at the same time keeping up the appearance of the sacrosanctity of the contracts, whose terms had only been renegotiated in order to restore the equitable division that both parties had originally intended.

4. Decline of the Fifty-Fifty Arrangement

With a fixed rent per barrel, landlords only needed to know the volume of production. With a percentage ad valorem royalty, they also needed to know prices. Thus far and no further went the involvement of private landowners in the USA.

However, in line with patterns set in the developed capitalist countries, the governments of petroleum exporting countries would inevitably adopt corporate income taxation. This type of tax generates very detailed information on the state of a business, which allows profit rates to be estimated and, therefore, the existence of excess profits. Indeed, between 1952 and 1961, the fifty-fifty resulted in a net profit rate for Aramco of 58%; the comparable figure for the Iraq Petroleum Company for the years 1952 to 1963 was 57%; for the International Consortium in Iran for the years 1955 to 1964 it was 69%; while in the case of the Kuwait Oil Company, for the years 1956 to 1960, it exceeded 150% (Mikdashi, 1966, pp. 181 and 221; and Mikdashi, 1972, p. 141).

The major international oil companies retained the most prolific oil acreage in the world, and paid for that privilege essentially the same rents and royalties, plus income taxes, as they would in marginal lands in the USA, which were the ones that determined international price levels. In terms of income taxation, these companies benefited from the fact that United States legislation considered oil fields to be deposits of valuable 'natural' commodities, as well as being co-owned by landlords and the companies that had explored for and discovered the fields. Therefore, both lessors and lessees, beyond the usual capitalist accounting, were allowed to deduct from their gross taxable income a depletion allowance, based on their gross income.

That is why the new concessions granted after the Second World War were significantly more advantageous for the respective states than the old concessions with their fifty-fifty arrangement. Notwithstanding the best efforts of the International Petroleum Cartel to retain all the petroleum acreage in Venezuela and the Middle East, in order to head off potential competitors at the pass, the fact is that plenty of land was left outside their control. In the new concessions, it became customary for there to be a state stake in the concessionaire company itself, over and above any fifty-fifty. For example, in

1957, the Italian state oil company Ente Nazionale de Idrocarburi agreed on a double fifty-fifty with Iran. That is to say that, on top of the usual fifty-fifty, Ente Nazionale de Idrocarburi offered Iran a shareholding participation of 50%. This would translate into a 75:25 profit split in favor of Iran. The American ambassador tried to intervene before the Shah to prevent such an arrangement from coming into being, but his entreaties were in vain. Indeed, in 1958, Iran granted a further two concessions under similar terms and conditions, one to a US-American company and another to a Canadian one.

4.1. Venezuela and the Middle East

In Venezuela, the fifty-fifty arrangement was called into question by the concession dealers. In the not too distant past – more precisely, up to 1938 – Venezuelan landlords had been given the opportunity of securing mineral concessions on their lands, with the objective that they could benefit from then selling them to the highest bidder. However, a few of these landlords, as well as selling the concession for a specified price, also managed to obtain an overriding royalty. Now, when the concessions in question were renewed pursuant to the 1943 Petroleum Reform, their rights vis-à-vis the concessionaires were preserved. Thus, overriding royalties would be subject to the Income Tax Law, and the petroleum sector rate of 12% was applied to them. But, subsequently, they were not subject to higher rates – pursuant to the extraordinary tax of 1945, and the 1946 reform of the Income Tax Law – which only applied to much higher incomes. However, overriding royalties were subject to the Additional Tax, created with the 1948 reform of the Income Tax Law.

For example, Henrique Otero Vizcarrondo and his heirs were paid a one half of one percent royalty for a concession in the eastern part of the country, and their income between 1948 and 1952 came to 1.6 million US$. The applicable income tax rate was 12%, so a total of 192 thousand dollars would be due in income tax. But the application of the Additional Tax would increase their tax liability by a further 608 thousand US$. They then appealed to the Supreme Court of Justice, questioning its constitutionality:

> The Additional Tax, as far as royalties are concerned, is not an income tax but a partial confiscation of capital which is not authorized by the Constitution […].
>
> A royalty is not a current income, whose source is being renewed. Rather, it is the form in which a capital is amortized […].
>
> A royalty is not an enrichment because, by the fact of being paid, the titleholder does not become richer than what he already was, but simply receives in successive periods something which already belonged to him and which lies underground.

> And since a royalty is not income, it should not be subject to a tax which is only levied on incomes (*Revista de Derecho y de Legislación*, p. 149f).

The Court did not enter into a discussion about the nature of royalty, but dismissed the petition given that petroleum reservoirs, in accordance with Venezuelan law, were the property of the Nation, and not of the landlords.

Acedo Payarez referred to this decision of the Court in the following terms:

> It would be a different matter if in Venezuela the owner of the soil were also the owner of the subsoil, as happens in other countries in which, for that reason, the legal situation of the one who exploits petroleum differs radically from that which they have according to the Venezuelan system. In the law of such countries, when the one who is exploiting petroleum is the landholder himself, every time that he sells any amount of hydrocarbons, he alienates a specific part of his patrimony [...]. His ownership [...] therefore diminishes with such operations as a function of, and in direct proportion to, the quantities extracted from the subsoil [...] (Acedo Payarez, 1972, p. 53; emphasize in the original).

Evidently, Acedo Payarez was alluding to the American depletion allowance, considering the companies who had discovered the deposits and the landlords as common owners.

Definitively, in Venezuela the owners of overriding royalties would be subject to an income tax rate of 50%, including the additional tax, whereas before they only had to pay at a rate of 12%. For the concessionaires, in contrast, the fifty-fifty arrangement only implied a modest increase in income tax, which stood at 28.5% before the application of the Additional Tax and increased – on average, and over the whole of the period when the Additional Tax was in force, that is, 1946 to 1957 – to only 31%. This reflected the fact that the surplus that would be subject to the 50/50 division was quantified by deducting not only production costs proper from gross income, but also rents and royalties. The latter were deemed to constitute part of the fiscal participation of the state in the 50% share of the benefits.

In reality, the fifty-fifty arrangement subjected the companies to an income tax rate which reflected the situation in the USA, which allowed them to deduct from their gross income the 'natural' value of their output. As we saw in Part One, the income tax rate generally applicable to commercial activities in the United States was 52%; but the effective rate for the production of hydrocarbons was reduced to between 37.7% and 26%. At a cost of production similar to those prevailing in Venezuela, the rate would have been around the middle of this interval; that is, around 32%. So this arrangement, essentially, implied that the companies no longer had to pay the corresponding income tax in the USA. However, United States legislation also allowed landlords a similar deduction on their royalty income, and the net taxable income was deemed to be 72.5% of the royalty.

In contrast, in the Venezuelan fifty-fifty arrangement, all rents and royalties were treated, altogether, as a share in the benefits. However, in Venezuela, "[t]he subsoil of the national territory [was] the absolute property of the Nation" (Acedo Payarez, 1972, p. 52; emphasize in the original), and the companies were not co-owners. Hence, in 1958, there arose a political discourse, with Salvador De La Plaza as its main spokesperson, which characterized the 1948 fifty-fifty arrangement as a fraud because a true fifty-fifty had to recognize the right of the landlord state, as sole owner of the natural resource, to a depletion allowance for the 'natural' value of the whole of the output, as well as a 50% share in the benefits determined pursuant to such a recognition (De La Plaza, 1959).

4.1.1. Fifty-Fifty and Sovereignty

The government of Rómulo Gallegos was toppled by a military coup on November 24, 1948, in which Lt. Colonel Marcos Pérez Jiménez took part. Subsequently, on December 2, 1952, General Pérez Jiménez would take office as President of the country. However, on the 23rd of January 1958, another civilian-military coup put an end to the government of Pérez Jiménez. The Government Junta immediately found itself facing an acute fiscal crisis; whereas the petroleum sector was buoyant, with a net profit rate, in 1957, of 32%, on all its Venezuelan investments, including the major investments in refining. Nevertheless, the President of the Junta, Rear Admiral Wolfgang Larrazábal, upon being sworn into office, made a commitment to the concessionaires that the existent legal arrangement would be fully respected, and that no changes would be made (World Petroleum, 02/1959, p. 41).

The Junta was supposed to be in power only long enough to organize democratic elections, which were scheduled for December 1958, whereupon power would be turned over to elected authorities. However, in November 1958, Larrazábal stood down from the Presidency of the Junta in order to participate as a candidate in the upcoming elections of December 7th. In his place there came in a lawyer and university professor, Edgar Sanabria, and, under his presidency, the Junta secretly prepared a reform to the Income Tax Law, on the basis of its extraordinary legal powers. The reform was announced over the radio and television on December 19, taking the entire country by surprise, the oil companies included. The Reform consisted in an increase to the highest income tax rate from 28.5% to 47.5%. With this change, the marginal share of the state in petroleum profits went from 50%, to 56.25% (= 1/6 + 47.5%*5/6).

This increase, in conformity with Venezuelan legislation, would apply to the year in course. Thus, in 1958, the Venezuelan government's receipts from petroleum reached the sum of 914 million US$, while the net profits of petroleum companies were 523 million US$. The split of the benefits, therefore, would work out at 64:36 in favor of the state.

In response to these developments, on December 22nd, the President of Creole, Harold W. Haight, sent a letter to the Minister of Mines and Hydrocarbons, Julio Diez, in which he expressed anger, surprise and alarm about the reform, and also requested that it be revoked, and threatened to take reprisals. That same day, Haight travelled to the United States for the Christmas holidays. At the airport, he made a statement before the assembled press:

> By this action, Venezuela becomes the first country in the world to break the so-called 50-50 principle of equal shares in the benefits of the industry, disregarding acquired rights and ignoring the moral if not legal obligation to negotiate this rupture with the interested parties (World Petroleum, 01/1959, p. 16; emphasize ours).

The Minister of Mines and Hydrocarbons responded immediately:

> [...] the tax reform [...] does not damage any vested right nor does it modify any existing agreement with the petroleum companies, since the so called 50:50 emanates from the law itself as opposed to a formal agreement on the matter.
>
> [...] the reform maintains, in its entirety, the contractual regime in the matter of concessions [...]. It is simply a modification of the tax regime, which [...] is of the sole and exclusive competence of national sovereignty. [...] Since this has nothing to do with contractual matters and is instead a sovereign act, the observations made by the President of Creole are baseless [...] (quoted in De La Plaza, 1959, p. 73).

Indeed, the reform did note even mention the additional tax. Landlords who benefitted from an overriding royalty would continue to be subject to the 50% Additional Tax, just the same as the concessionaires. However, for the concessionaires the Additional Tax would now be zero. Haight, for his part, would not return to Venezuela.

4.1.2. Fifty-Fifty and the Depletion Allowance

The president of Creole insisted on the Additional Tax of 1948, and pretended that vested rights were recognized to the concessionaires connected to it. But the Sanabria government displayed the tax sovereignty of the country, which had been formally acknowledged by the concessionaires as a result of the 1943 Petroleum Reform. In support of its actions, the government made no mention to any intrinsic value that the mineral resource might have.

The situation was different in all the other petroleum exporting countries, where both income tax and the additional tax were contractual and all disputes that could arise in regard to the concessions would be subject to international arbitration. Tax sovereignty was not on the political agenda for any one of these countries. For all of them, it still came down to negotiating and renegotiating different aspects of existing arrangements. Presenting oil deposits as stores of valuable 'natural' commodities was a powerful ideological tool, of practical importance at the moment of negotiating a new arrangement, a true fifty-fifty, predicated on correcting an alleged conceptual error. Moreover, the concessionaires would be unable simply to ignore this position in question, as within the USA all of them supported the idea that petroleum deposits did indeed constitute stores of valuable 'natural' commodities. Hence, a 'true' fifty-fifty would actually consist of two elements: compensation for the 'intrinsic' value of the resource plus a fifty-fifty split of the remaining benefits. This 'intrinsic' value – in line with a properly Ricardian approach – was identified with the one-eighth royalty, the customary royalty rate in all the other petroleum exporting countries. The idea was to achieve an increase on the state marginal share of 50% by taking it to a one eighth royalty plus 50% of the remaining benefits, which would also translate into a share of 56.25% (= 1/8 + 50%*7/8).

4.1.3. Conclusions

The fifty-fifty arrangement had arisen in Venezuela in 1948, on the initiative of the US-American oil multinationals, in an effort to refuse to the state, de facto, its sovereign rights regarding the international petroleum rent, even though these rights were anchored, de jure, in the 1943 Hydrocarbons Law. In all the other petroleum exporting countries, such tax sovereignty was also denied de jure. With the Sanabria Decree, Venezuela was the first country to put de facto an end to the fifty-fifty arrangement, exercising its sovereign rights. But even though Venezuela was the first petroleum exporter to take such a step, it did so prudently, picturing all of the rents and royalties collected beyond the royalty of one eight as differential rents, which made it possible for the other petroleum exporting countries to follow its example and, as part of a common front, thereby increasing the minimum state share in the benefits from 50% to 56.25%, even if this goal had to be achieved through negotiations and taking as its basis the ideological construct of the 'intrinsic' value of petroleum.

4.2. Venezuela: The Hemispheric Treaty

The Sanabria Decree coincided in time with the fact that the international petroleum price structure was fracturing into two, a major development that would overshadow all other problems which would come up in the ongoing development of the international petroleum regime.

While the USA had been a net exporter, petroleum prices all over the world were equivalent to domestic US-American prices, plus the cost – whether real or deemed – of shipping from an US-American port to the place in question. Once the USA became an oil importer in 1947, prices the world over would be set by reference to American domestic prices minus the shipping costs – whether real or deemed – from the respective place to an US-American port. Even so, with the high domestic price level in the USA – where prices had doubled with the end of the Second World War – and the continuous development of productivity in maritime transportation, the resulting prices were still very high in comparison with the production costs of the petroleum exporting countries.

This structure endured thanks to the domestic control of production by the Interstate Oil and Gas Compact Commission, on the one hand, and an import quota system administered by the International Petroleum Cartel, on the other hand. This arrangement was bolstered by the informal but effective cooperation of the United States federal government. From 1953 onwards, the producing provinces of Canada and Venezuela participated in the Commission in question as observers. In Canada, production was subject to control by provincial governments, the same as in the USA, but in Venezuela, production quotas were established by agreements among the major producing companies.

In 1958, this informal arrangement entered into crisis due to the progressive weakening of the International Petroleum Cartel, on account of the output coming from the new concessions in the Middle East and, in particular, very recently, in Venezuela. The rising imports put downward pressure on domestic US-American prices, threatening the viability of innumerable marginal wells. The United States government then put into place a system of obligatory import quotas, centered on refineries, which would thereby have access to petroleum from the international market, at lower prices, and would pocket the corresponding extraordinary profit which, moreover, would now become taxable income in the USA, and no longer in Venezuela.

Hence the USA put an end to the privileged access to its domestic market for Venezuelan petroleum, obviously in retaliation to the Sanabria Decree, commonly known as Hemispheric Treaty, but continued to maintain it for Canadian petroleum. Faced by this situation, the new Venezuelan government quickly set up the Coordinating Commission for the Conservation and Commerce of Hydrocarbons (Comisión Coordinadora para la Conservación y el Comercio de Hidrocarburos) (CCCCH), which would have the faculty to deny export permits to cargos whose sales prices were deemed too low in comparison to domestic US-American prices. But this was supposed to be only a first step, as the government explained to the National Assembly on April 11, 1959. The idea was that Venezuela would:

> [...] foster the formation of an international organism which regulates petroleum prices worldwide, just as there is one for coffee, sugar and other raw materials (Quoted by Melcher, 2010).

The idea, then, was to set up an intergovernmental agreement for a basic staple. Moreover, the haste of the Venezuelan government in setting up the CCCCH was also related to the First Arab Petroleum Congress, which was to take place in Cairo towards the end of that same month of April.

4.3. Gentlemen's Agreement and Sovereignty

In 1949, the Venezuelan diplomatic delegation which travelled to the Middle East visited Iran, Kuwait, Iraq and Egypt. Wherever it went it extended invitations to the First National Petroleum Convention, which was to be held in Caracas in 1951. Farman Farmaian attended on behalf of Iran, but also in attendance was Abdullah Tariki in representation of Saudi Arabia, a country which only two years earlier had denied visas to the members of that Delegation. Subsequently, the Arab League prepared the first Arab Petroleum Congress, which would finally take place in Cairo, in April 1959. Both Venezuela and Iran would be invited to attend as observers.

As a matter of fact, Tariki had returned to Venezuela in 1956, but this time as Director General of the Petroleum and Mining Directorate of the Ministry of Finance and Economy of Saudi Arabia. During this visit, he made the acquaintance of an American attorney, Frank Hendryx, who was the General Counsel of Creole (SONJ). Hendryx had come to the conclusion that Creole's policy of denying Venezuela its sovereign rights, a denial inherent to the fifty-fifty deal and, in general, SONJ's formal denial of such sovereign rights to the other petroleum exporting countries, subjecting them to international arbitration, was wrong and would, sooner rather than later, end up by failing. In late 1958, Hendryx quit his post at Creole, and Tariki would retain him as an advisor.

When the Betancourt government took office, Juan Pablo Pérez Alfonzo was again entrusted with the petroleum portfolio, and as the head of the Ministry of Mines and Hydrocarbons, he was to lead the Venezuelan delegation that would attend the First Arab Petroleum Congress. Pérez Alfonzo had just spent ten years in exile and he was not personally acquainted either with Farman Farmaian or with Abdullah Tariki, who headed the Iranian and Saudi delegations, respectively. But Manuel R. Egaña, who was also part of the Venezuelan delegation, did know them personally, and took it upon himself to make the relevant introductions. Pérez Alfonzo then took the initiative to strike a confidential deal with them and the representative of

Kuwait, amongst other participants, to be known as the Gentlemen's Agreement. The signatories made the commitment to put before their respective governments the proposal to set up a Consultative Petroleum Commission, as well as reaching a consensus on some specific points:

> The consensus of opinion was that the governments [...] should be guided by a 60:40 formula as a minimum to bring themselves into line with the recent position assumed by Venezuela and the tendency in new contracts in other countries (Acosta Hermoso, 1969, p. 18).

However, Pérez Alfonzo introduced into the discussion a further topic that was close to his own heart. In Venezuela, government take depended, as far as income tax was concerned, on market prices and, by means of the CCCCH, the Venezuelan government was determined to arrest their continued slide. In the Middle East, in contrast, income tax was based on posted prices. Therefore, for example, when Aramco reduced in February 1959 its posted price for its reference crude oil – the Arab Light 36o API – from 2.08 US$/b to 1.90 US$/b, the consequence was a corresponding fall in the fiscal income of the Saudi government. However, the signatories to the Gentlemen's Agreement had reached consensus that any future change to prices "shall be discussed beforehand and be approved by all interested parties". The signatories also agreed about:

> The need to set up in each country organs that will coordinate, from the national standpoint, the Conservation, Production and Exploitation of Petroleum (Acosta Hermoso, 1969, p. 18).

Regarding the Congress itself, Frank Hendryx delivered an address on behalf of Saudi Arabia, entitled A Sovereign Nation's Legal Ability to Make and Abide by a Petroleum Concession Contract (Hendryx, 28/04/1959), in which he praised the notable advances that Venezuela had made on this matter by means of sovereign action. Furthermore, he stressed that, under international law as well as under American, English and French national law, sovereign nations had the right and the obligation to modify concession contracts if the public interest so demanded.

5. Organization of the Petroleum Exporting Countries

During the following months, Pérez Alfonzo and Tariki were very active in the promotion of the idea of an international organization that would regulate the global production of petroleum and, by extension, of petroleum prices. Even though they encountered many problems, Pérez Alfonzo was confident of a successful outcome, as evidenced by a letter that he sent Egaña on August 22, 1960: "Ultimately, we will get to, it cannot be doubted, an International

Compact or Commodity Agreement" (quoted by Melcher, 2010; emphasize in the original).

5.1. International Prorationing

On August 9th, 1960, Aramco once again cut the reference price for Arab Light, from 1.90 US$/b to 1.80 US$/b. Post haste, those responsible for the petroleum policy of Saudi Arabia, Iraq, Iran, Kuwait and Venezuela met in Baghdad during the month of September 1960 and established the Organization of the Petroleum Exporting Countries (OPEC). Its Resolution I.1 expressed the need for there to be agreement between all the interests involved in the international petroleum market, both on the side of the producing nations as well as the consumers, in pursuit of the twin objectives of a fair return for the petroleum industry and stable prices. To this end:

> [...] Members shall study and formulate a system to ensure the stabilization of prices by, among other means, the regulation of production.

During the II Conference, in January 1961, in Resolution II.13, OPEC stated that "having made a tentative study of the prevailing prices generally", it would undertake "a detailed study [...] in order to arrive at a just pricing formula, supported by a study of international proration, should that prove essential". And in Resolution II.14, it made an overture to the US-American government:

> [...] having discussed the matter of quotas and other restrictions imposed on the imports of petroleum by friendly countries in order to protect their domestic production and reserves, expresses the hope that these countries will consult with the Members [...] regarding such quotas or restrictions in order to arrive at satisfactory solutions and thus promote mutual understanding for the protection of the interests of both exporting and importing countries.

Venezuela, for its part, was already preparing a United States-style production control and, in the Third OPEC Conference, it highlighted once again the successes achieved by the CCCCH. However, the international petroleum companies responded to the possible participation by exporting countries in the international pro-rationing of production with a stark 'no' and, for good measure, threatened the Middle East countries with international arbitration in the event that the latter decided to follow the Venezuelan example. With the possible exception of Iraq, none of these countries was ready to risk such a confrontation. According to Ashraf Lutfi, Secretary General of OPEC between 1965 and 1966, the companies' stark 'no' was explained by the following:

> The majors' ability to manipulate production in the Middle East has been, and still is, a potent weapon, which they have used and will no doubt continue to

> use to very good effect against individual countries in the area. Agreement on their part to operate in accordance with an OPEC Joint Production Program would effectively entail a relinquishment of the weapon that has enabled them to surmount many a storm in the past – for the production program is in essence an instrument whereby OPEC itself, rather than the oil companies, can assume the responsibility for deciding on the production level from the OPEC area as a whole, as well as the output from each member country (Lutfi, 1968, p. 68).

5.2. Proprietorial Rent

At the IV Conference, in June 1962, in the face of total silence from the developed countries, Resolution IV.32 recommended the immediate opening of

> […] negotiations with the oil companies concerned […], with a view to ensuring that oil produced in Member Countries shall be paid for on the basis of posted prices not lower than those which applied prior to August 1960.

The idea would no longer be that market prices be reflected in posted prices. Instead, the latter would be converted into fiscal reference prices, so that downward movements of market prices would no longer have an impact on fiscal petroleum income. Or, to put it another way, the idea was to manage the fiscal income per barrel produced, and not the market price. Price reductions would only hit the concessionaires' bottom lines. Moreover, Resolution IV.33 proposed – on the basis of the fiscal reference prices – to account for the royalty as a cost reflecting the "compensation for the intrinsic value of petroleum".

The international petroleum companies, for their part, refused even to talk to OPEC. The companies would only hold discussions with, for example, Secretary General Fouad Rouhani, in his capacity as representative of Iran. Furthermore, they offered different countries, on an individual basis, some specific advantages like, for instance, a faster increase in production. While the companies had not reversed the price cut of August 1960, it was also true that they had refrained from any further price cuts. So their offer of a privileged increase in production, coupled with the conversion of the posted prices of August 1960 into frozen fiscal reference prices, was an interesting offer per se.

Moreover, in their negotiations with the countries in question, the companies had no option but to accept, in principle, the accounting of the royalty as a cost – reflecting the 'intrinsic value' of the mineral resource – even if only on a partial basis. All future progress on this proposal would be conditioned on a recovery in market prices. Among the countries directly affected, four – Saudi Arabia, Iran, Kuwait and Libya – were willing to accept the company proposals but one – Iraq – was not, taking OPEC to the brink of dissolution. The crisis was only overcome in November 1964, when the VII

Conference took place, and member countries agreed to disagree on the matter.

OPEC had been forced by the turn of events to find its own way. Ultimately, its response to lower prices had to be the creation of a proprietorial rent. In this endeavor, it was to be strikingly successful. Definitely, rent per barrel increased slightly.

5.3. Declaratory Statement of Petroleum Policy in Member Countries

After leaving behind its internal divisions in November 1964, OPEC recovered and forged ahead, in the context of an irresistible worldwide process of decolonization. Of particular importance in this regard were the Arab-Israeli conflict as well as the revolutions in Arab countries, which faced a variety of different political structures, both monarchical and republican. After all, the majority of OPEC's members were Arab countries. However, all OPEC member countries were united by their national interest – i.e. not right wing or left wing, neither monarchic nor republican, but quintessentially national – and this national interest was reflected in leveraging the value of the national ownership of the natural resource on the international stage.

The Third Arab-Israeli War of 1967 was, without a doubt, the triggering event for OPEC's subsequent progress. In 1968, it approved Resolution XVI.90, entitled Declaratory Statement of Petroleum Policy in Member Countries, the most important of all its resolutions. This policy was summarized in a series of Principles, of which the most relevant are the following.

Referring to the policy of Participation in the shareholding capital of the Concessionaires, it stated that the Government "may acquire a reasonable participation, on the grounds of the principle of changing Circumstances". The main spokesman for this policy was Saudi Arabia's Minister for Petroleum Ahmed Zaki Yamani, who favored participation and opposed the policy of nationalization. Yamani argued that with such participation, the issue of production control would also be solved: member countries would be able to restore equilibrium to the market, if the need arose, on the basis of their volume entitlements derived from their shareholding in the concessions. Thus, in the countries of the Middle East, an increasing participation in the capital structure of the concessionaires would represent an evolutionary path towards sovereignty. The starting point would be the reference to a 20% government shareholding in the Iraq Petroleum Company concession title deeds.

The Resolution did not refer to tax sovereignty, but only to Posted Prices or Reference Prices, to be determined by the Government. Governments would have the faculty of granting a Limited Guarantee of Fiscal Stability "for a reasonable period of time", but this would in any event be subject to a

Renegotiation Clause, because operators "shall not have the right to obtain excessively high net earnings".

The Resolution also dealt with the Settlement of Disputes. In principle, "all disputes arising between the Government and operators shall fall exclusively within the jurisdiction of the competent national courts or the specialized regional courts, as and when established". Evidently, those member countries which lacked the necessary legal infrastructure and were not in a position to develop one anytime soon, would be able to resort to arbitration in specialized regional courts to be created.

Finally, as regards the Mode of Development of their natural resources:

> Member Governments shall endeavor [...] to explore for and develop their hydrocarbon resources directly. [...]
>
> [...] when a Member Government is not capable of developing its hydrocarbon resources directly, it may enter into contracts of various types [...] with outside operators for a reasonable remuneration [...] (Emphasize ours).

Note that, if necessary, the governments would resort to foreign operators.

The Policy Declaration did not dwell on the 'sovereignty' as such of the member countries, but only on their 'permanent sovereignty'. In its introductory part, it made reference to "the inalienable right of all countries to exercise permanent sovereignty over their natural resources", as "a universally recognized principle of public law", one furthermore "repeatedly reaffirmed by the General Assembly of the United Nations". Permanent sovereignty would amount to a non-extinguishable and ongoing right to re-negotiate any arrangement.

With this Policy Declaration, the gap between Venezuela, on the one hand, and all other member countries, on the other, narrowed significantly. Venezuela, in fact, continued to be the reference par excellence for all other member countries. In the words of Ashraf Lutfi:

> In my references to OPEC's problems about the need for financial adjustments and revision of the abnormal legal provisions in oil agreements, I should, of course, have made exception for Venezuela on many grounds. Many of the problems discussed are the exclusive concern of the Middle East Countries since they have been to a large extent solved in Venezuela. There it is the government that to some degree plays the paternal role vis-a-vis the oil companies, and, on the whole, oil operations in that country are not favored with special privileges unavailable to other commercial and industrial sectors. The oil operations, foreign and local alike, are subject to the law of the land and pay whatever taxes the government sees fit to impose generally on all sectors without discrimination. If disputes arise between oil operators and governmental agencies, such disputes are taken to the local courts as a matter of course. [...] the Venezuelan government is able to recommend any level of

income taxation without having to obtain the prior approval of the foreign operators, as is the case in the Middle East. In general, therefore, the Venezuelan Government is free from any fetters on its authority to formulate the oil policy of the nation regarding existing oil exploitation arrangements (Lutfi, 1968, p. 72).

5.4. Conclusion

In August 1960, the posted price of Arab Light was 1.80 US$/b. With a production cost estimate of 0.20 US$/b, the fifty-fifty split left the fiscal participation of Saudi Arabia and the net profit for the concessionaires at 0.80 US$/b. At the beginning of 1970, that same price continued to hold as the fiscal reference price, but in the meantime the market price had fallen to 1.25 US$/b and production costs had shrunk to 0.12 US$/b. But Saudi Arabia's government take had increased to 0.85 US$/b; meaning that the net profit for the companies decreased to 0.28 US$/b. The profit split was now 75:25 in favor of the government.

From the point of view of the petroleum exporting countries, fiscal reference prices reflected what they considered to be the 'true' market prices, based on domestic US-American prices. But from the standpoint of the concessionaires, the fact of the matter was that Saudi Arabia levied a 50% tax on a non-existent income, calculated on the basis of a fiscal reference price of 1.80 US$/b, when the prevailing market price was only 1.25 US$/b.

6. The Decline of the American Reference

In the USA, the average wellhead price of a barrel of oil in 1959 was 2.90 US$; ten years later, in 1969, it had increased to 3.09 US$. In the meantime, over that same interval, the price of an imported barrel of oil in the Federal Republic of Germany had decreased from 3.00 US$ to 2.25 US$. This price differential translated into a loss in the competitive capabilities of US-American industry, since the prices of petroleum were the main driver of the cost of energy in general. In line with this difference, the USA witnessed an increase in the political tensions between the interests of petroleum producers and those of consumers, which reached a critical point towards the end of the decade. The American consumers insistently demanded lower prices, on the basis of an increase in petroleum imports. At the beginning of 1970, a commission named by President Nixon did indeed recommend the adoption of such a policy.

But as a matter of fact, the USA was already on the verge of a crisis in the hydrocarbons sector. That same year, the Department of State concluded that US-American domestic prices would not go down to the levels prevailing in the international petroleum market; instead, the reverse would happen: prices in the international market would rise to the levels prevailing in the domestic

US-American market. It was estimated that, by 1980, the price of a Persian Gulf barrel would reach 4.50 US$ (Akins, 04/1973, p. 464).

The American government was taken completely by surprise by the fact that, in 1970, petroleum production in the country reached its historical peak. It was not an unforeseeable event as such. In fact, in 1956, a renowned geophysicist in the employ of Shell, M. King Hubbert, predicted that this peak would be reached around the year 1970. However, the companies were totally uninterested in such predictions, which would only strengthen the position of the landlords in the negotiations for new leases. The companies would always claim that they had access to sufficient – indeed, even abundant – reserves and that they were therefore not really interested in acquiring new leases. This explains why both the United States government and public opinion in general and, probably, the international petroleum companies themselves, overestimated their production potential.

6.1. Economic Sciences and the Price of Petroleum

As regards the world petroleum market, neither United States public opinion nor, in general, that of any other developed capitalist country, paid any attention whatsoever to OPEC. Furthermore, all the economists who were experts in the energy field were of the unanimous opinion, supported by modern Economic Sciences, that OPEC was irrelevant as far as market prices were concerned. In the past, such prices had been determined on the basis of domestic prices in the USA, which meant that the extraordinary surpluses in the exporting countries were conceptualized as international differential rents. In consequence, when free access to the domestic United States market was restricted, international prices would necessarily have to fall on account of competition.

The most prominent among these experts, Morris Adelman (professor of economics at the Massachusetts Institute of Technology), emphatically stated in 1964:

> The landlord, as residual claimant, simply gets what is left, much or little. […] The whole problem of rents and royalties is superfluous to the determination of price (Adelman, 1964, p. 108; emphasize in the original).

Adelman advised both sides to the ongoing negotiations that any arrangement would have to be flexible enough to adjust without complications as prices continued to fall, which, in his view, was inevitable. His advice was completely ignored by the major international oil companies, as well as by the landlord States. As things turned out, the ground rent per barrel not only stabilized, but actually increased slightly. But Adelman had an explanation for this phenomenon: this increase only reflected the fact that exporting countries had still not appropriated all available differential rents by means of the fifty-fifty.

Adelman entrenched himself in his theoretical position and, at the beginning of 1970, he reasserted his prediction of a continuous fall in petroleum prices for the next fifteen years, on the basis of studies that confirmed decreasing costs to discover and develop an incremental petroleum barrel worldwide. He had no doubt that, under pressure from the forces of competition in the market, the international petroleum companies and the exporting countries alike would finally be forced to adjust to lower rents and royalties.

Starting from the assumption of competition in the market as the sole relevant factor – i.e. ignoring the existence of OPEC – the 1970 estimates of the Department of State – i.e. a price of 4.50 US$/b in the Persian Gulf by 1980 – was possibly a very realistic estimate. But suddenly OPEC and its members turned up as actors, and actors of first importance at that. And these actors were not only going to appropriate the whole of the extraordinary surpluses created by the rise in prices but, in addition, they would push prices to levels considered unimaginable until then, with a consequent explosive rise in the ground rent per barrel.

In the USA, and in developed countries in general, there arose then fantastic theories that sought to account for the turn of events by suggesting a conspiracy between the international oil companies and their respective governments, on the one hand, with OPEC, on the other, since the latter had suddenly emerged into the limelight like a rabbit from a magician's hat. Among those promoting this kind of speculation was none other than Adelman himself, who refused to admit that his theoretical scheme might be somewhat deficient. But facts are stubborn and in 1972, Adelman finally admitted:

> The per-barrel tax is now the basic support of the current price level, and many times more important than cost or company profits (Adelman, 1972, p. 207).

Now, all of a sudden, Adelman referred to the fiscal regime and the per barrel tax as the floor for market prices. This did not lead him to reconsider his theoretical standpoint; instead, he observed that, as a practical matter:

> [...] we must consider what forces might breach even a strong floor (Adelman, 1972, p. 211).

It is worth highlighting once again the place that Ricardian rent theory occupies within the Economic Sciences. The theory is presented as a model of reality but, when it clashes with reality, it is reality that is seen as being in the wrong. Ricardian rent theory is, without doubt, a political program: it cannot be admitted that the tenure of land might have any incidence on prices.

6.2. Conclusions

Adelman reached the conclusion that the international concessionary regime no longer allowed the companies to defend the interests of the petroleum importing countries, even in conditions of 'perfect competition'. It was imperative to put an end to it, and to transform the international companies into associates, contractors or service providers of the new national petroleum companies, with a view to making sure that the latter companies imposed their will on their home governments, thereby breaking the 'floor' as a way of improving their competitiveness (Adelman, 1974).

Between 1920 and 1973, world crude oil production grew on average at a 6.6% annual rate. Up until the Second World War, USA petroleum production represented about two thirds of the world total, but this percentage quickly shrank thereafter and, by 1970, the United States share in world oil output had decreased to a mere fifth. Even more, in 1970, oil production in the USA – which at the time was still the largest individual producer – reached its historical peak, at 9.6 million b/d. By 1973, this had shrunk to 9.2 million b/d, a sixth of the world total. American fields, after more than a century of intense exploitation, were becoming ever more depleted but, since domestic demand kept growing, net hydrocarbon imports had to expand at a very quick pace to make up the difference. From 1947 to 1970, imports went from zero to 2.8 million b/d; but by 1973 they already came to 5.7 million b/d. Conversely, petroleum production by OPEC member countries began to account for an ever-growing share of the world total, going from 41.5% in 1960, to 51% by 1970, and 55.7% in 1973 (Table II.1).

Table II.1 CRUDE OIL PRODUCTION 1920 – 1973				
Year	USA (%)	OPEC* (%)	Others (%)	Total (TB/D)
1920	64.4	4.4	31.3	1,882
1940	62.9	15.8	21.3	5,874
1950	51.9	32.3	15.8	10,419
1960	33.6	41.5	24.9	20,920
1970	21.1	51.0	27.9	45,654
1973	16.5	55.7	27.8	55,657

* Membership of 1970: Algeria, Indonesia, Iran, Iraq, Kuwait, Libya, Nigeria, Qatar, Saudi Arabia, United Arab Emirates and Venezuela.

Source: Venezuela, Ministry of Mines and Hydrocarbons: *Petróleo y otros datos estadísticos*, several years.

Early in 1973, the United States government froze domestic petroleum prices, as had been done throughout both the Second World War and the Korean War. On those two occasions, the measure had been enough to freeze international oil prices in their turn but this time the companies were no longer in a position to contain them. Furthermore, during 1974, spare production capacity (a constant feature of the United States petroleum industry since 1949) disappeared, and the government put an end to import quotas. United States domestic petroleum prices would no longer determine international petroleum prices. Now, the reverse would be true.

One would have thought that it was incumbent upon the government to evaluate the potential, at both the domestic and the international level, of a natural resource with the undeniable strategic importance of hydrocarbons. But in the United States, up until 1973, no such assessments featured in official statistical compilations, due to opposition from the lessee companies. This situation would only change after 1973 and, from 1977 onwards, a newly formed Department of Energy would be entrusted with compiling and analyzing all relevant information.

7. International Petroleum Crisis

One of the consequences of the Third Arab-Israeli War of 1967 was the closure of the Suez Canal for a period of more than six years. This had the effect of further improving the already advantageous geographical situation of Algeria and Libya relative to the European petroleum market. Both of these countries had become petroleum exporters before joining OPEC, in the midst of a market in recession. The international oil companies had set relatively low posted prices, and maintained these prices even after the closure of the Canal. Hence, towards the end of the decade, these two countries put it to the concessionaires that they should also benefit from OPEC's successful initiative in terms of fiscal reference prices and, that in addition, such prices should also be adjusted in accordance to the differential rent that reflected their advantageous geographical location.

The disagreement on the matter reached its climax in 1970, when Libya imposed a production cut on the basis of its petroleum conservation legislation, which had been drafted with the assistance of OPEC. France attempted to hold the line in Algeria by imposing an embargo on its former colony, but was unable to obtain any support from the USA. Equally unfruitful was the proposal put to the United States government by Royal Dutch-Shell that Libya be invaded. Ultimately, Libya and Algeria would prevail in this dispute, with OPEC's open support and assistance. Libya had joined OPEC in 1962, while Algeria only did so in 1969.

In 1970, the world petroleum market went from being in surplus to a situation of tightness, and extreme tightness at that, since petroleum production peaked simultaneously in the USA, Venezuela and Libya. It would only be possible to compensate the fall in supplies from these countries with incremental production from the Middle East which, furthermore, would also have to satisfy the rising global demand.

Between 1969 and 1970, there were a number of price increases in the United States domestic market, which in total came to around 0.45 US$/b. These were the first price increases to occur in twelve years. In November 1970, under pressure from the Iranian government, the companies accepted an increase of 0.09 US$/b in the fiscal reference price for Iranian Heavy crude oil, as well as an increase to 55% in the income tax rate. These benefits were extended immediately to Kuwait and Saudi Arabia.

This radical change in market circumstances took OPEC, the oil companies and all petroleum importing countries alike completely by surprise, but OPEC happened to be well positioned to take full advantage of it. At OPEC's December 1970 Conference, which took place in Venezuela, the member countries agreed that, from January 1971, royalties would be expensed in full rather than partially, before the application of income tax, whose minimum rate was to be increased to 55%, while fiscal reference prices would be uniformly increased in all member countries.

The Conference had not yet finished its deliberations when the Venezuelan Congress approved a reform to the Income Tax Law which not only increased the rate applicable to petroleum companies uniformly to 60%. At the same time, the National Executive was given the right to set, unilaterally and by decree, the fiscal reference prices for Venezuelan petroleum.

The negotiations with the Persian Gulf countries were scheduled to begin in brief in Tehran. Moreover, OPEC agreed to hold an Extraordinary Conference in Tehran to monitor the progress of these negotiations and, if necessary, to take unilateral measures.

In the meantime, the United States government hastily put together and coordinated a united front of 22 petroleum companies with a view towards the negotiations in Tehran, suspending the applicable anti-trust legislation. Obviously, in these negotiations, the companies were to represent the interests of consumers, who were the ones that would bear the economic consequences of any agreed increase in international ground rent. In other words, what was being discussed was nothing less than an increase in proprietorial rent.

The companies ultimately caved in before all of OPEC's demands. The agreements provided that the fiscal reference price of Arab Light would increase progressively over the next five years (i.e. by 1975) from 1.80 US$/b – the prevailing price since August 1960 – to 2.62 US$/b, which implied that ground rent would increase from 0.91 US$/b, in December 1970, to 1.53 US$/b in 1975. Fiscal reference prices for Libya and Algeria were to be adjusted accordingly.

7.1. Participation in the Capital of the Old Concessionaires

As far as the policy of participation in the shareholding capital of concessionaires was concerned, Algeria, Libya and Iraq had already taken a series of unilateral measures, prompted by the Third Arab-Israeli War. After the Conference of December 1970 in Caracas, OPEC as a whole began to move in pursuit of this goal. The key protagonist in these negotiations, as representative of the Arab coastal countries of the Persian Gulf, was Sheikh Ahmed Zaki Yamani, Minister of Petroleum of Saudi Arabia. In December 1972, Yamani announced that he had reached an agreement with the international petroleum companies in question. Starting in January 1973, the Arab states of the Persian Gulf would obtain a shareholding participation of 25% in all the old concessions, which was to progressively increase up to 51% by January 1983.

For his part, the Shah of Iran announced in March 1973 the dissolution of the International Consortium, and the assumption of all operations by the National Iranian Oil Company. The erstwhile members of the Consortium would thereafter only act as service companies, and the National Iranian Oil Company would sign 20-year supply agreements with them (Stobaugh, 1978, p. 221).

7.2. Climax of the International Petroleum Crisis

The Tehran Agreement was a mere respite for importing countries and the international oil companies. When the dollar was devalued in 1971 and, once again, in February 1973, OPEC responded with commensurate increases in fiscal reference prices, in order to preserve the purchasing power of its members. But the subsequent price spike would surpass all expectations. In 1973, demand for OPEC petroleum rose by 14.4% and the increase in market prices outstripped the agreed fiscal reference prices. OPEC recalled the oil companies in order to renegotiate the Tehran Agreements.

On October 6th, 1973, Egypt and Syria surprisingly unleashed the fourth Arab-Israeli war and on October 15th, the Organization of Arab Petroleum Exporting Countries (OAPEC) – which had been formed after the Third Arab-Israeli war, and encompassed all Arab petroleum exporting countries (whether or not they were also members of OPEC) agreed to implement a series of production cuts, as well as a selective embargo directed principally against the USA and The Netherlands. Then, on October 16, in Kuwait, the OPEC ministerial committee grouping the Persian Gulf countries announced that OPEC would break off negotiations, and would instead pursue unilateral sovereign actions. The fiscal reference price of Arab Light was increased from 3.01 US$/b to 5.13 US$/b, which would mean that ground rent per barrel would rise from approximately 1.77 US$ to 3.05 US$.

7.3. The End of the International Concessionary Regime

The production cuts by Arab countries – which reached as much as 20% of output – provoked panic in the market and forced the consuming countries to adopt radical saving measures such as, for example, the prohibition to use automobiles on a Sunday.

The Ministerial Committee met again towards the end of December in Tehran, with representatives from all the other member countries participating as observers. There, it was decided to increase the fiscal reference price of the Arab Light benchmark to 11.65 US$/b; taking the ground rent to 7.00 US$/b.

The fiscal reference price was kept unchanged at 11.65 US$/b throughout 1974, but Saudi Arabia increased the royalty rate to 20% and the income tax rate to 85%. By the end of 1974, the government take – including the share in profits arising from participation – had reached ten dollars per barrel.

Fiscal reference prices would disappear before long, to be replaced by posted prices once again (by the end of the 1960s, posted prices were 40% lower than fiscal reference prices), which in turn were meant to reflect market prices. Furthermore, the posted price for Arab Light would serve as the keystone for the whole international price structure. At any given location, the price of a barrel of oil would be determined by taking this price and adding the transportation cost – whether deemed or real – from the Persian Gulf to the location in question, apart from any necessary quality adjustments.

Moreover, Saudi participation in Aramco was increased, retroactive to January 1973, from 25% to 60%. By 1983, this was to reach 100%.

7.4. Conclusions

In 1970, the OPEC member countries benefited from an amount of 7.6 billion US$ in international ground rent; by 1974, this had grown tenfold to 70 billion US$ (in 1970 money). OPEC had taken advantage of the favorable juncture to impose its Policy Declaration. But then, with the fourth Arab-Israeli war and the selective OAPEC embargo as triggering events, OPEC assumed the mantle of determining fiscal reference prices on a sovereign basis and also set a target for a majority participation in the shareholding capital of the concessionaires. The petroleum exporting countries, as sovereigns and shareholders, would henceforth freely determine both prices and volumes.

The evolutionary development of the international concessionary regime, with a new stage centered on mixed enterprises, was scarcely able to reach an embryonic stage and came to an end in 1974. The developed capitalist petroleum-importing countries had just suffered a major defeat, and were forced to retreat. The concessionary regime had come to its end.

But, inevitably, the developed countries would return to the offensive, on the basis of a new petroleum regime still to be created, which would be in accordance with the condition inherent to the development of capitalism; namely, the subordination of land to capital. In the Tehran negotiations of 1971, the international petroleum companies, on account of the emergency, directly represented the interests of consumers. Nevertheless, the fact is that such a role only corresponded to them to the extent that they competed among themselves, within the framework of an appropriate regime. That is to say, every company would seek to minimize, to the extent possible, all the rents, royalties and taxes to be paid to resource owners, in the same way as they would minimize all other costs in order to maximize their profits, but free competition would ensure that the benefit of their efforts would finally go to consumers. Creating a new global regime would require, for certain, the intervention of the governments of the developed capitalist and petroleum-importing countries.

Part III.

Globalization and Petroleum Fiscal Regimes

The first dizzying rise in oil prices happened between 1973 and 1974 and, a few years later, between 1978 and 1980, there was a second rise, even more significant, due to the Iranian revolution and the subsequent war between Iran and Iraq. Theses rises had an adverse impact on the Gross Domestic Product (GDP) of developed capitalist countries. The International Monetary Fund) estimated that the impact came to 2.6% of GDP in the first case, and 3.7% of GDP in the second case (International Monetary Fund, 08/12/2000). Of course, rising prices would also drive down demand. Between 1960 and 1973, production of crude oil had grown at an average annual rate of 7.8% but, between 1973 and 2017, it would only do so at an average rate of 2.5% (Figure III.1).

Figure III.1

WORLD CRUDE OIL PRODUCTION AND OPEC
1960 – 2017

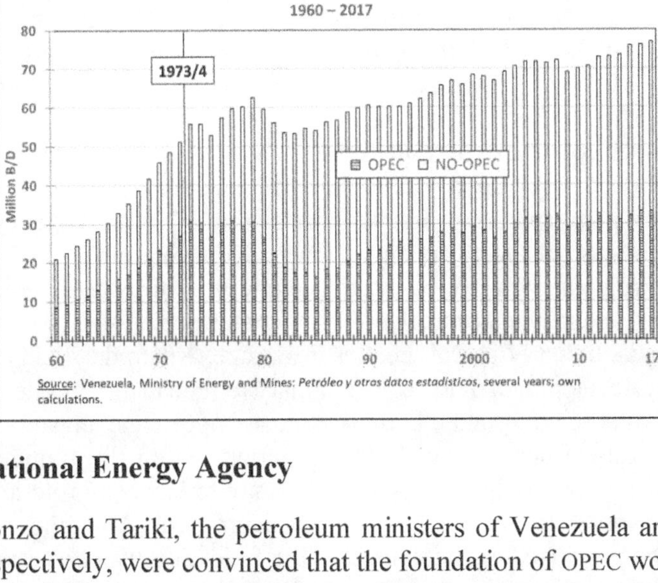

Source: Venezuela, Ministry of Energy and Mines: *Petróleo y otros datos estadísticos*, several years; own calculations.

1. International Energy Agency

Pérez Alfonzo and Tariki, the petroleum ministers of Venezuela and Saudi Arabia, respectively, were convinced that the foundation of OPEC would only be a first step in the way to an intergovernmental agreement – a *commodity agreement* – between the major consuming and petroleum exporting countries. Neither expected a categorical *no* on the part of the former as an answer to such an initiative.

Tariki was to be succeeded by Yamani, who promoted a policy of *Participation* in the share capital of the incumbent concessionaires, an option which seemed to open an evolutionary path which, in essence, would have the same goal as the commodity agreement: stabilizing market prices. However, this *Participation* policy had barely seen the light of day when, in 1973, there arose the confrontation between the Arab petroleum exporting countries and developed capitalist countries, triggered by the fourth Arab-Israeli War. On 17 October 1973, OAPEC announced that its members would reduce their output by 5% each month, until all developed capitalist countries changed their posture relative to the Arab-Israeli conflict. Furthermore, in the following days, individual member countries also announced a selective embargo, mainly targeting the USA and the Netherlands.

OPEC, for its part, took advantage of this confrontation in order to impose the principles it had enunciated in 1968 in its *Declaratory Statement of Petroleum Policy of the Member Countries*. On the other side, the consuming countries and, therefore, their international oil companies, would implacably square up to OPEC, denouncing this organization as a 'cartel' of landlord States.

The United States government reacted to the Arab petroleum embargo by setting up, in 1974, the International Energy Agency (IEA), within the framework of the Organization for Economic Cooperation and Development (OECD), made up of the developed capitalist countries. Set up in 1961, the OECD was committed to promoting economic development and global commerce. It was headquartered in Paris, where the IEA was also to be located.

Even though OPAPEC formally ended its petroleum embargo in August 1974, the IEA nevertheless adopted immediately an International Energy Program whose objective was the collective tackling of future emergencies. The member countries committed themselves to store enough petroleum to cover, if necessary, up to sixty days of consumption. After the second vertiginous price spike, towards the end of the 1970s, the Agency agreed to increase these stocks to cover ninety days of consumption.

The Program was to be managed by the *private* international oil companies. In the new political circumstances, national state companies were no longer particularly useful, neither in Europe, where they had appeared as part of the development of the concessionary regime in the Middle East, which had just collapsed, nor in the Latin American importing countries, where they had been established during the same period within the framework of an import substitution policy. They were also not to have any role in terms of a national response to any international petroleum crisis. All of these companies were to be privatized in the course of the following twenty years, with one important exception: Norway retained its state company Statoil, created in 1972. Furthermore, this country only took up a conditional membership in the IEA.

Quite apart from the International Energy Program, the IEA member countries would invest large sums in promoting the domestic production of hydrocarbons, as well as alternative sources of energy. They would only purchase a barrel of crude oil and products originating in OPEC when there was no alternative. In 1973, the OPEC member countries produced 30.6 million b/d of crude oil and the rest of the world 25.1 million b/d; that is to say, OPEC's share of total world crude oil production came to 55%. By 1985, OPEC's production had decreased to 16.2 million b/d and petroleum production in the rest of the world had increased to 37.8 million b/d; OPEC's share of world crude oil production had shrunk to 30%. Finally, OPEC – and in particular Saudi Arabia – decided not to continue reducing its own production to defend high oil prices, so the latter eventually collapsed. This collapse can be appreciated in Figure III.2, which plots the average prices for the Venezuelan export basket of crude oil and products, as well as the recovery in the call on OPEC crude oil, which would eventually stabilize at between 40% and 45% of world petroleum demand.

Figure III.2

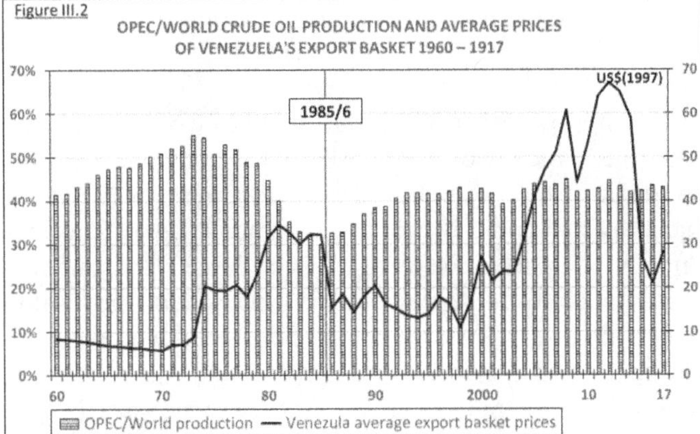

Sources: Figures III.1 and V.5.

2. Non-Proprietorial Fiscal Regime

Faced with a second dizzying price increase, which materialized towards the end of the 1970s, the IEA recommended that its member countries should take all possible measures "to minimize declines in their own indigenous petroleum production", and also to revisit their respective processes for granting licenses or concessions, as well as their fiscal regimes, "so as to encourage timely development" of their domestic resources (Scott, p. 169). The idea was to eliminate bidding bonuses and royalties – both of which could affect negatively the viability of a petroleum project – thereby transforming their fiscal regimes into non-proprietorial ones, that is, with a marginal ground rent equal to zero. Moreover, at the end of their term, concessions or licenses were to be extended as a matter of course and without there being any bonus payments. Hence, any vestiges of a proprietorial fiscal regime would be eliminated[1].

Moreover, thanks to the development of technology, it was possible to drill at ever greater water depths, even under exceptionally adverse climate

[1] A non-proprietorial fiscal regime was what Adelman had recommended that the parties go for when the *fifty-fifty* split was re-negotiated at the beginning of the 1960s, precisely to avoid putting the brakes on the downward movement of oil prices (which he, in any event, considered inevitable).

conditions. Thus, in 1968 petroleum was discovered in the North Sea, which would become the most important new petroleum province subsequent to the 1973 'OPEC Revolution'. The North Sea petroleum province fell essentially under the jurisdiction of Norway and Great Britain, and it was here that the new reference for petroleum regimes would develop, to supersede and substitute the erstwhile United States reference which had until then underpinned the international concessionary regime.

2.1. United States of America

However, the non-proprietorial fiscal regime would have to confront a proprietorial one, which was profoundly rooted in the most important country within the ranks of the IEA: The USA. Furthermore, with leases in the USA being subject to fixed and pre-arranged contractual conditions, both landlords and producing companies were to obtain thereto unimaginable benefits. Consumers as such were represented by the federal government, but the private ownership of minerals was, politically speaking, beyond questioning. As a consequence, as far as the federal government was concerned, it all came down to minimizing the market impact of such ownership. Let us turn to two examples of this process.

2.1.1. Alaska

In Alaska, due to its natural conditions, public ownership of land prevailed. When, in 1969, the state organized its first bidding round for petroleum leases in the North Slope, it did so on the basis of a one eighth royalty rate and a 3% severance tax. The sole bidding parameter was a cash bonus. To widespread astonishment, bonus payments came to 900 million US$. This result was the first signal that hydrocarbons production in the lower 48 states of the USA was about to reach its historical peak.

Subsequent Alaskan bidding rounds, from 1973 onwards, were based upon a royalty rate of one sixth, and one fifth after 1982, always with a cash bonus as sole bidding parameter. The severance tax for petroleum was raised concurrently, up to a value of 15% by 1982.

Moreover, in the USA, the states corporate income taxes are generally determined as a percentage of the nation-wide income for federal corporate income tax purposes, which makes their calculation straightforward. In Alaska, the applicable percentage was 9,4%. However, in 1978, Alaska ring-fenced the oil companies and insisted that the sum total of the companies' extraordinary profits had to be reflected in the quantification of their taxable income for state purposes; they could not be diluted in the accounts for corporate income taxation at the federal level.

In 1970, Alaska's population came to barely 300 thousand inhabitants. Quite abruptly, the government ramped up public spending and attempted to promote new investment projects. However, it soon became evident that the capital absorption capacity of the state was very limited. Indeed, the capital absorption capacity of a country, in general terms, is proportional to productivity: to produce is to invest, and to invest is to produce. However, the immense bonus payments bore no relationship with the productivity of the state's economy. Thus, any attempt to invest beyond this limit, whether the funds involved were public or private, was destined to fail.

Therefore, in 1976, Alaska took the decision to create a Permanent Investment Fund – to be endowed with a significant portion of the state's yearly petroleum rent – which was to be invested *outside* the state; that is to say, in the other states of the Union or else abroad. Thus, the state's population would obtain an annual benefit in the form of a dividend and, at the same time, the Fund would be a legacy for future generations when the oil and gas fields were exhausted. The decision to set up the Fund, in fact, was put to a vote and, having been approved by a qualified majority, it was incorporated into the Constitution of the state.

In 1977, the initial endowment of the Fund was 734 thousand US$, and by 2019, this figure had grown to 64 billion US$. With population having grown in that interval to 730 thousand inhabitants, the Fund represents a capital of 88 thousand US$ per Alaskan.

Fiscal Regime

However, already by 1977, the oil companies had convinced the Alaskan government that it was necessary to make the severance tax more flexible, reducing it to zero as a field neared the end of its life. And in 1981, the government was forced by litigation promoted by the oil companies to put an end to the fiscal ring fence, and to return to the usual accounting derived from federal income taxation.

Moreover, in the USA, leases awarded through bidding rounds cannot be renegotiated. Consequently, the only way that a leaseholder has to change the terms of a lease is to return it, with a view to acquiring it again at a subsequent auction in more favorable conditions. Nevertheless, in 1995, the government of Alaska gave the Commissioner for Natural Resources ample powers to negotiate lower royalty rates in state leases, "to allow for production that would not otherwise be economically feasible" (Alaska, 1995, §38.05.180), and also to adopt sliding scale royalties according to prices. A wave of lease renegotiations ensued.

With these changes, Alaska's fiscal regime drew ever closer to a non-proprietorial one (B. Mommer, 2002, Chap. 7.2).

2.1.2. Federal Lands

In all of the other states of the Union, public lands were surrounded by private landholdings and it was therefore impossible to attempt to eliminate even signature bonuses, let alone royalties. However, starting in 1981, the Federal government began to put up for auction regularly *all* available land (as opposed to selected blocks), thereby swamping the market and thus bringing bonuses down significantly. Of course, this had an incidence over both public and privately held lands. Then, from 1995 onwards, the Federal government authorized the *Minerals Management Service* to exempt from the payment of royalties – albeit for a limited time only – certain Federal offshore deep-water leases (Boué, 2007).

2.2. Norway

Norway granted its first petroleum licenses towards the end of the 1960s. These had a term of 46 years, and an extension of 500 km^2, which had to be reduced by half at the end of nine years. In line with the United States reference, the royalty was one eighth. From the very first discoveries, it became clear that Norway would become a petroleum exporting country for all the foreseeable future (Figure III.3).

Figure III.3 NORWAY: PRODUCTION AND EXPORTS OF HYDROCARBONS 1976 – 2015

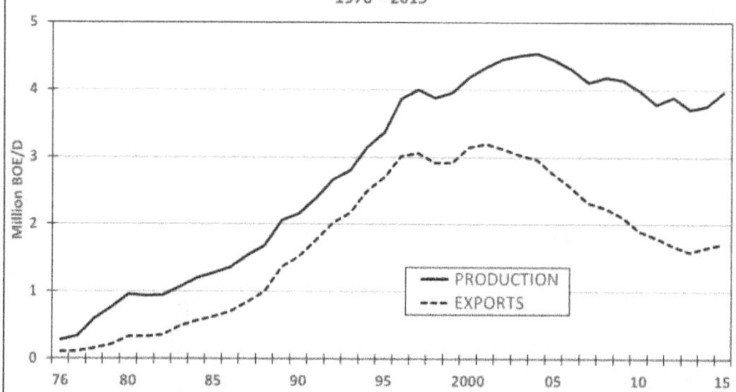

BOE: Barrels of Oil Equivalent

Source: www.norskpetroleum.no

Norway was a developed country, but had no indigenous petroleum industry, a situation which its government now considered unacceptable. For this reason, in 1972, the government established Statoil as a state oil company, which was to have a majority shareholding in all licenses. However, with the increases in both Norwegian exports and international oil prices, already by the mid-1980s Statoil had acquired immense power, due to its control over outsize cash flows. The government therefore took the decision to transform it into an operating company that would have a minority shareholding in the consortia that held the licenses. Statoil would therefore be constrained to manage the projects through agreements with its private partners. The balance of Statoil's shareholding in the licenses was transformed into a State Direct Financial Interest, which generated a cash flow that would go directly to the government's coffers. The voting rights of the State Direct Financial Interest would be exercised by the Norwegian Petroleum Directorate. The Directorate was also entrusted with a regulatory role in all technical matters, including oversight and monitoring of the prices to ensure that the company would sell oil at market prices. For its part, the Ministry of Petroleum would be in charge of legislative matters and initiatives, defining petroleum policy, planning and, moreover, organizing licensing rounds.

In 1986, and in line with the IEA recommendations, Norway eliminated royalties altogether and, at the same time, proportionate adjustments were made to the State Direct Financial Interest. Finally, production licenses were to be extended as a matter of course at the end of their term, without any bonus payment.

Norway would also decrease its corporate income tax rate in line with what the USA did during the presidency of Ronald Reagan. At the same time, however, it increased the rate of the so-called supplementary income tax that only applies to the petroleum sector. Thus, the sum of income taxes of this sector would always remain at around 65%.

Each participant in a production license was required to file their tax returns separately. For such purposes, corporations that participated in a number of licenses were allowed to report their upstream income on a consolidated basis, subject to a ring-fence drawn at the *national* level. Thus, when Statoil began to expand internationally from the 1990s onwards, it had to do so on the basis of a strict separation between its Norwegian financials, on the one hand, and its international financials, on the other. Thus, national extraordinary surpluses would not thereby be diluted by means of *global* consolidated accounting.

In 2001, Statoil was partially privatized, with a view to forcing it to compete with the efficiency criteria of private enterprise. However, the state retained a shareholding participation of 67% in the company. At the same time, the management of the State Direct Financial Interest was maintained completely independent from Statoil, by means of the creation of *Petoro*, a company which would be the titleholder of the shares of the state in the licenses (Al-Kasim, 2006).

2.2.1. International Investment Fund

Facing a surge of international petroleum rent, Norway, for her part, had to face the limits of her capital absorption capacity. Thus, in 1990, following the example of Alaska, Norway created an Investment Fund destined for investments *abroad*. A very significant part of the international petroleum rent was to be channeled to the Fund. The Fund addressed Norway's inability to absorb oil rents in times of boom, and its yields would provide income in times of low prices or upon depletion of reservoirs.

In 2020, the value of this Investment Fund was estimated at 1.4 trillion US$. Given Norway's population of barely 5.2 million inhabitants, this represented a capital of 269 thousand US$ per inhabitant.

2.2.2. Conclusions

The Norwegian fiscal regime is non-proprietorial and, formally, it presents itself as such. Prudently, Norway always held its distance vis-à-vis OPEC and its price policies. Norway also maintained domestic prices of derivatives of hydrocarbons at international levels, as among other things to discourage domestic consumption by having in place excise taxes comparable to those customary throughout Europe. More generally, the country always took part in, and even promoted, a policy of sustainable development, which discourage excessive consumption of natural resources (whether renewable or not) and contributed to protection of the environment.

As the only developed oil country, Norway designed its petroleum regime in accordance with the global development of capitalism, on the basis of market mechanisms and free competition, and without any distinction being made between the domestic market and the export market. Thus, it was never in doubt that Norway would dedicate its international ground rent to *investment,* even though, inevitably and necessarily, it would have to invest *abroad*.

Of course, a non-proprietorial fiscal regime would entail a high level of taxation for the domestic part of the petroleum sector also, as well as an equally high level of excise taxation on the consumption of petroleum products. However, these taxes on the national petroleum sector could be compensated with, for example, a corresponding reduction in taxes on the non-petroleum sector. Definitely, from the standpoint of the domestic market, the natural resource was a free gift of nature, and not a source of ground rent.

3. Global Fiscal Regime

Nevertheless, even within the framework of a non-proprietorial fiscal regime, the exploitation of a given field will be abandoned when the increase in unit production costs – due to progressive depletion – exceeds the unit revenues obtainable at market prices. When this point is reached, however, due to the lower cost of production in earlier years (when it generated extraordinary surpluses), the *average* cost of production of the depleted field during its lifetime will still be lower than the unit cost which prompted its abandonment. But regarding the extraordinary surpluses of the past, whether they were captured by the state and then distributed nationally, or whether they had been appropriated by the producing companies, this would change nothing as far as the abandonment of the field was concerned.

However, the situation would be very different if the state intervened to capture the extraordinary surpluses in question in the form of a *contingent liability*, that is, a debt payable to the same producing companies, to be amortized when the unit cost of production began to exceed the unit revenue obtainable at market prices, in a way that would allow the company exploiting the depleting field to continue obtaining the customary profit rate. Companies would therefore continue exploiting depleting fields, up until the point when the balance of the contingent liability reached zero. Thus, a first period during which the state would capture extraordinary surpluses would be followed by a second period where these captured surpluses would be used to boost returns diminished by depletion up to the customary profit rate. Hence, the *average* unit production cost over the whole period would equal the average unit revenue at market prices, which would entail the disappearance of all extraordinary surpluses, both at a *national* and at an *international* level. Definitively, this would be a *global* fiscal regime, as it would convert the natural resource into a free gift of nature for the global consumer. Under such a regime, petroleum producing countries, whether they were net importers or net exporters, would benefit only from the expansion and spillover effects of their respective domestic petroleum industries.

3.1. Great Britain

Great Britain became a producer of hydrocarbons during the mid-1970s, and a net exporter between 1980 and 2005 (Figure III.4). Nevertheless, the country had an extraordinarily rich historical experience in petroleum policies, structured around two companies whose origins were international, and finally global: British Petroleum and Royal Dutch-Shell[2]. Moreover, Great Britain had always looked with suspicion upon the United States reference which, nonetheless (and very much to the regret of the British), had ended up becoming universal. But when the international concessionary regime collapsed, Great Britain returned to the stage as a new and important producing country as well as an exporting country, even if it would only belong to the latter camp for twenty-five years. Furthermore, not only was the natural resource itself under public ownership, it was also offshore and literally out of sight, which meant that the country was therefore ideally situated to create the new reference for a *global* petroleum regime.

---

[2] The Royal Dutch Petroleum Company arose, in 1890, in Indonesia, at the time a Dutch colony. Shell Transport and Trading Company was a British company, founded in 1897. In 1907, they merged to become Royal Dutch-Shell, a company whose share capital is 60% Dutch and 40% British.

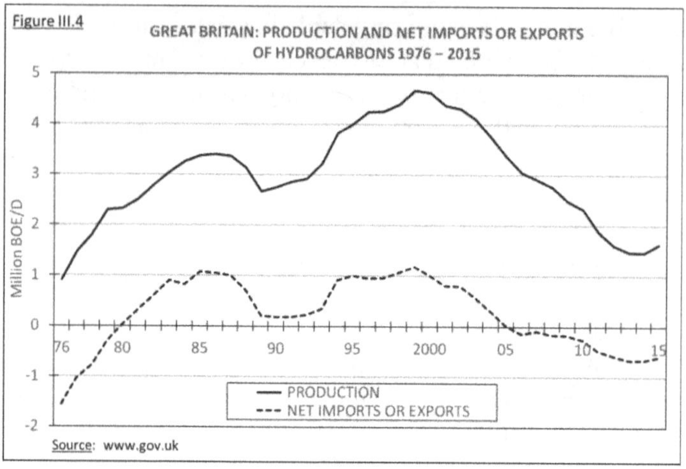

The first UK petroleum licenses – the UK does not use the term 'concession' – of 250 km^2 were granted towards the end of the 1960s, for a period of forty years. They were assigned on the basis of negotiations about a minimum and legally binding work program. The customary royalty rate was one eighth.

However, subsequent to the first significant discoveries and the sustained increase in international oil prices during the first half of the 1970s, the British government set itself the task of reforming the petroleum fiscal regime in order to turn it not only into a *non-proprietorial* regime, but also into a *global* one.

3.1.1. Contingent Debt: First Phase

In 1975, Great Britain created the Petroleum Revenue Tax (PRT). Strictly speaking, this levy was not a 'tax' so much as a *contingent liability* based upon the extraordinary petroleum surpluses. This 'tax' would be applied to cumulative cash flow, and would treat capital investments and current expenses in the same way, with annual adjustments according to the customary rates of profit, interests and inflation. In any given field, the PRT would only be due when the cumulative cash flow was positive. It would be applied to revenues generated during the respective fiscal year (which in Great Britain begins on April 6th). Definitely, the PRT, as a levy on excess profits, would have no incidence on the production cost of the marginal barrel. Even more, if the cumulative cash flow were to decrease, for whatever cause[3] (and

[3] After the *Piper Alpha* catastrophe of 1988, in which 167 workers died, the affected companies enjoyed negative Petroleum Revenue Tax payments for a number of years, due to the losses suffered. Furthermore, all companies active in the North Sea were forced to comply with new and costly safety requirements. As a result, in 1990, total Petroleum Revenue Tax payments were *negative*, to the tune of 216 million UK£.

especially as the field reached its economic production limit), the companies would be reimbursed from the PRT funds paid in previous years[4].

The Petroleum Revenue Tax was to be applied to each field individually, and it would be subject to a ring fence which sought to prevent the dilution of extraordinary surpluses with costs from activities such as transportation, refining or distribution and, also, to prevent extraordinary surpluses from being artificially reduced by transfer prices or the sub-contracting of all kinds of activities.

Determining the Petroleum Revenue Tax rate was the prerogative of the Executive, which initially set it at 45%. Towards the end of the 1970s, when oil prices rose once again by leaps and bounds, the rate reached a maximum of 75%, but with the subsequent price weakening it was reduced once again, this time to 50%. Furthermore, the PRT was abolished for fields yet to be developed because these were considered, *a priori*, to be marginal. These fields would simply pay the usual taxes levied on the non-petroleum sector.

3.1.2. Contingent Debt: Second Phase

UK hydrocarbons production began to fall in 1999, and the descending trend continued despite the very high subsequent prices. In order to contain the decline, large additional investments would be required – in infrastructure too – and their profitability would be guaranteed by means of the contingent liability.

Since 1998, the British government was already discussing with the industry the manner in which the second phase of hydrocarbon exploitation in the North Sea was to be managed; that is to say, how to succeed in maximizing the economic recovery of remaining reserves. The answer, first of all, was to leave behind the approach followed during the phase of the Petroleum Revenue Tax inflows, which was based on individual fields. Liquidating the contingent liability on this basis would certainly optimize recovery of individual fields but would not optimize output on a province-wide basis. To achieve the latter, a great deal of cooperation between the various producing interests involved would be required. The major part of the infrastructure had been created to serve the large fields that had come on stream thirty years before and were approaching obsolescence. The option of renewing it would not always be profitable, unless it were possible somehow to take into account the numerous small fields which depended on that infrastructure, but which had never paid any PRT. For this to happen, the creditors of the contingent liability would be allowed to invest these sums within the relevant area, beyond the boundaries of the specific individual field.

---

[4] Nevertheless, *all* publications, without exception, refer to the Petroleum Revenue Tax simply as a 'tax', and not as a 'contingent liability'.

In 1999, output reached its historical peak of 4.7 million barrels of oil equivalent per day, and by 2015 it had decreased to 1.6 million. According to studies undertaken by the government, after a cumulative production of 42 billion barrels of oil equivalent, remaining recoverable reserves would be between 12 and 24 billion. Had the creditors of this contingent liability not been allowed to invest it beyond the boundaries of individual Petroleum Revenue Tax-paying fields, the recoverable volume would be around 12 billion barrels of oil equivalent; whereas by allowing it, the recovery expectation for the next three decades was a volume close to 24 billion.

For the corresponding arrangement, the Oil and Gas Agency was created in 2016, with the status of a 'government company', invested with ample powers to promote and enforce the required cooperation but also to come to agreements with industry. The rate for the Petroleum Revenue Tax was set at *zero*, effective 1 January 2016, so from that point on, the PRT cash flows would only be negative (i.e. the contingent liability would be liquidated). Furthermore, the discretionary faculties of the government with regard to changes in the PRT rate were extinguished by an act of Parliament, so future changes to this rate could only happen pursuant to an act of Parliament.

In this way, the phase of liquidation of the contingent liability to creditor companies kicked off, with the specific aim not only of increasing the recovery factor of specific fields, but of the whole province.

3.1.3 Non-proprietorial Fiscal Regime

From the very beginning, Great Britain aimed at a non-proprietorial fiscal regime. But given the violent increase in international prices towards the end of the 1970s, Parliament enacted a Supplementary Petroleum Duty – effectively a severance tax, which itself is a royalty of sorts – at a rate of 20% for fiscal years 1981 and 1982. However, subsequently, there began the process of gradual elimination of the one eighth royalty, which disappeared for good in 2002, as the first licenses were about to reach their full term, and which would be renewed under a non-proprietorial fiscal regime.

Up to 1983, the corporate income tax rate was 52%, in line with the United States reference on the issue. Subsequently, Great Britain would follow the downwards path of the USA corporate income tax, but would then continue to reduce it all the way down to 20% by 2016. In any event, in 2002, simultaneously with the elimination of the royalty, the executive had created a supplementary charge on income tax at a rate of 10%, which would only apply to the petroleum sector and which the government was to manage at its discretion. With the subsequent price increases, the rate reached a high point of 32% in 2011, but it followed prices on the way down thereafter, and by 2016 was once again at 10%. Furthermore, with the creation of the Oil and Gas Agency, the discretionary faculties of the government with regard to the supplementary charge also came to an end.

3.2. Conclusions

As a practical matter, the Petroleum Revenue Tax was designed with extreme care in order not to affect investment under any circumstance. As the Figure III.5 shows, from the early 1980s onwards, there was a decreasing trend in fiscal income per barrel, despite an overall rising trend in prices. Finally, in 2015, income taxes amounted to 538 million UK£ but the PRT was a negative 562 million UK£. Thus, the fiscal contribution of North Sea hydrocarbons production to the British budget was actually *negative*, an unprecedented occurrence.

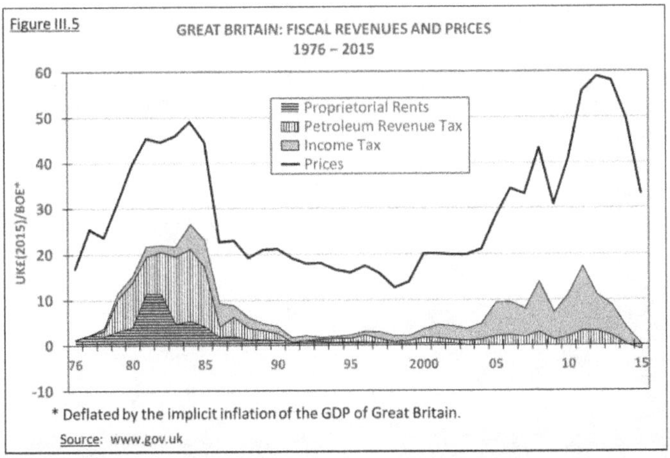

Of course, just like Norway, Great Britain kept domestic prices of hydrocarbons at international levels, and also kept excise taxes high, thereby discouraging domestic consumption and, therefore, the demand for OPEC oil.

In 2015, the contingent liability came to 133 billion UK£; assuming a population of 65.1 million for Great Britain as a whole, this contingent debt worked out at two thousand pounds sterling per inhabitant.

4. Global petroleum regime and the nation state

The British petroleum regime was conceived to be essentially a global one, with the aspiration of becoming the new international and national reference, for importing as well as for exporting countries, that would replace the US-American reference of old. It was the product of an eminently *national* policy, but of a country with a *global* vision, implemented by Labour and Conservative governments alike, in the context of a confrontation with OPEC. In terms of its design, this regime was underpinned by the global experience of British Petroleum, Royal Dutch-Shell, and Exxon.

A global fiscal regime takes aim at both proprietorial as well as differential rents. In essence, the global fiscal regime is predicated upon the assessment of extraordinary surpluses over the life of a field and, ultimately, of a whole producing region, which means that these rents go down to zero. Of course, upon expiry of any license, renewal for it under the non-proprietorial regime would be granted as a matter of course, and the corresponding contingent liability would be kept on the books.

Norway, for its part, merely implemented a non-proprietorial fiscal regime. Lacking an indigenous petroleum sector, Norway created one *ex nihilo*, but always leaving room for foreign capital. Furthermore, with its international Investment Fund, it became an important *creditor* country. Hence, Norway never had to face a crisis arising from international indebtedness and it was therefore never vulnerable to pressure on the part of international financial institutions to transform its national petroleum regime into a global one.

Part IV.

Venezuela: The Petroleum Regime

Throughout the first three parts of this work we have dealt with the topic of petroleum from the perspective of developed capitalist countries. We will now continue focusing on it from the standpoint of exporting countries, taking Venezuela as an example.

Venezuela was the most important petroleum exporting country throughout the international concessionary period, as well as the most advanced among its peers in both legal and political terms. Furthermore, because of its geographical location and early independence, it did not have to face the complexity of the national and international political circumstances which surrounded the other petroleum exporting countries, and which tend to obscure the analysis of the petroleum question.

1. Fixed Duration and Reversion

At the beginning of the 20th century, Venezuela's mining legislation did not attribute any value to the natural resource, which meant that concessionaires were expected to pay taxes equivalent to those faced by other enterprises, while paying nothing for the right to exploit the country's hydrocarbons resources. Thus, in the Exposition of Motives of the 1909 Mining Code, it was pointed out that, while the concessions would be granted for a maximum term of fifty years, they could be renewed so that their holder, "pressured by the expiry of the term provided for in the contract, will not try to destroy or waste the mineral resources found therein, in order to extract from the mine, the greatest output in the least time possible". The state granted concessions for the sole purpose of attracting foreign investment.

Hence, if a concession were to revert for any cause, the state would not indemnify the concessionaire for any improvements or permanent fixtures. Thus, in 1918, in the Regulations for Coal, Petroleum and Similar Substances (*Reglamento de Carbón, Petróleo y Sustancias Similares*), it was set forth that, upon the expiry of a concession, "the Government shall not be obligated to pay for improvements of any kind, but shall receive the mine with all of its buildings, machinery and attached works" (art. 17).

In 1917, as soon as Venezuela became an exporting country and petroleum a source of international ground rent, there developed a hydrocarbons-specific legislation. Nevertheless, as far as reversion was concerned, the 1943 Hydrocarbons Law maintained the position that:

> [...] the Nation shall re-acquire, without the payment of any indemnity whatsoever, the parcels that had been granted, and shall likewise become the owner of all the permanent works that may have been constructed thereon (Art. 80).

The concessions would have a term of 40 years as a maximum but, at the request of the concessionaires, between the 20th and 38th year, and for one time only, the concessions could be extended for a further forty years. However, it was specified that in the extension agreement:

> The conditions [...] shall be those which the Federal Executive may consider advisable to agree upon with the concessionaire; but in no case are they to be less favorable to the Nation than those governing the concession at the time of the agreement.

In general terms, pursuant to the 1943 Hydrocarbons Law, the Venezuelan petroleum regime drew level with that prevailing in lands under the jurisdiction of the US Federal government, but the limited term of the concessions marked an important difference. In fact, in the Exposition of Motives, it was pointed out that:

> [...] mining legislation universally sets forth the right to exploitation not by defined terms but until the exhaustion of the mineral, and this is recognized in the United States of America with respect to hydrocarbons.

The 1947 Constitution, for its part, specifically provided that the Executive could not grant a concession for an undefined term (art. 138).

However, petroleum as a source of international ground rent became ever more important to the national economy, and therefore reversion as defined in the 1943 Hydrocarbons Law no longer made any sense. The issue was not whether one concession would revert to the state, but rather that all the concessions granted between 1943 and 1944 would do so at the same time. These concessions were much more important than those granted in 1956 and 1957. However, when in 1959 the international petroleum market weakened, some of the international oil companies began to dismantle their exploration rigs and ancillary equipment in order to re-export them. The Ministry – with Pérez Alfonzo as Minister – attempted to make the exporting of the equipment in question a matter that required the approval of the Ministry, arguing that such equipment would be subject to reversion, which meant that the state would be acquiring a property right in this equipment over a period of forty years, at an annual rate of 2.5%. The government thus interpreted in this broad manner the reversion clause, seeking to appropriate the whole of the investment of the concessionaires on a *national* level, and including both real property and chattels. But Gulf and Shell took the government to the Supreme Court of Justice, where their position prevailed. The Court held that reversion:

> [...] is an expectation of a right in favor of the Nation, which only materializes at the end of the term of the concession and which extends only to the permanent works affixed at that point to the conceded areas and thus does not affect the right which concessionaires have, throughout the term of the concession, to carry out in such areas whatever construction works, modifications and demolitions which are most convenient in terms of the exploiting interests (*Gaceta Oficial de Venezuela – GOV*, 26/12/1963).

In view of this judgment, it would appear as if the situation were comparable to the one prevailing in the USA towards the end of the 19th century, when similar verdicts ultimately resulted in the term of petroleum leases being made co-extensive with the exhaustion of the discovered fields. But the Venezuelan petroleum regime was subject to the hegemonic power of the state, both as a sovereign and as a *landlord*. For now, the government withdrew to await more favorable times. Indeed, the world petroleum market put the wind back in its sails and, in August 1971, the Venezuelan Congress enacted the Law on Goods Subject to Reversion in the Hydrocarbon Concessions (*Ley sobre Bienes Afectos a Reversión en las Concesiones de Hidrocarburos*), to which we will refer to as Reversion Law. The basis of this law was the conception of a petroleum concession being a public service – a thesis which Pérez Alfonzo had advanced already in 1943, in his *Dissenting Vote* – and which therefore obligated the concessionaire to guarantee the continuity of the service granted, and also to maintain all the productive structure in a fit and proper state.

Moreover, the companies continued to hold concessions over a total of 2.9 million hectares, out of which only around 20% had been explored. The Reversion Law provided that unexplored areas were to revert to the state within a period of three years, and when all the other areas eventually reverted, the entirety of the concessionaires' assets within Venezuela would become the property of the state, 'without any compensation whatever being due'.

The companies went to the Supreme Court once again, requesting that the Reversion Law be declared unconstitutional because of its confiscatory character. But this time, in 1974, the Court held in favor of the state.

2. Production Control

In June 1954, when the Pérez Jiménez government was about to put into practice its intention to grant new petroleum concessions, Pérez Alfonzo, then in exile, sent the President a telegram in which he warned that this action would be "a cause of major damage to Venezuela", because "competition among producers would not be desirable [...]. Such competition would translate into decreasing prices contrary to the direct interest of the nation, as a participant in the profits" (Pérez Alfonzo, 1961, p. 82f).

Betancourt likewise was of the opinion that:

> [...] Venezuela, a producing country on a grand scale, has no interest in a war for market shares among operating companies, but in international agreement and regulation (Betancourt, 1956, p. 797).

Betancourt had no doubt that the exporting countries would have to belong to some kind of international agreement and regulatory scheme:

> A future popular government will have plenty of arguments with which to justify the intervention of the state to determine – together with the industry,

> but in a position of pre-eminence – the paths of production. Such arguments are to be found, by the dozens, in the US government's report on the Petroleum Cartel. From that report emerges the evidence that seven gigantic enterprises decide, years in advance, what will be the extraction quota of crude of each producing area, the refining volume of each one of them [...]. Also fixed with anticipation, in their general terms, are the market prices for the mineral. And with regard to these procedures, two observations are in order. The first one, is that the very existence of the Cartel as an international regulator of the industry in its production and sales aspects appears in the current circumstances as the sole obstacle against a fierce war for markets between the major trusts, when what is desirable is that, in future, a global agreement between governments might be reached to rationalize production across all continents [...]. The second one, for all that the cartelization of the petroleum industry is a consequence of a process that is already very much advanced, nonetheless it is unacceptable that this gigantic financial combine functions as a conspiracy between 'seven majors'. It is inadmissible [...] that the states which represent nations that are owners of fields exploited under concession systems have no say when long-term production plans are drawn up [...]. The representatives of the producing nations must have highly rated seat, voice and vote in the industry's planning bodies (Betancourt, 1956, p. 886).

When Betancourt took office as president once again in February 1959, with Pérez Alfonzo as his Minister of Mines and Hydrocarbons, his fears in terms of the consequences of the granting of new concessions by Pérez Jiménez were being confirmed. Post-haste, the new government created then the Coordinating Commission for the Conservation and Commerce of Hydrocarbons (*Comisión Coordinadora para la Conservación y el Comercio de Hidrocarburos* – CCCCH), whose mission was conceived, first and foremost, as a means to prevent the new oil companies, with their recently acquired concessions, from forcing their way into the US-American market through discounts. The new government hoped to recover the 'Hemispheric Treaty' for Venezuelan petroleum, which had recently been suspended. However, the whole situation had also become considerably more complicated by the Sanabria Decree because the Betancourt government was definitely not willing to revoke it.

At the same time, during the First Arab Petroleum Congress, held in Cairo in April 1959, the government used the CCCCH to promote the idea of an 'International Compact' for the control of production, which would include the most important producing countries, exporters as well as importers.

3. Fiscal Regime

Even with the increase in petroleum income arising from the Sanabria decree, the Betancourt government was nonetheless forced to shrink public spending. As a consequence, in 1960, it reduced the salaries of public sector workers and devalued the bolívar from 3.09 to 4.40 Bs/US$. However, the 3.09 Bs/US$

exchange rate was preserved for oil company transactions, and this bespoke exchange rate would only be devalued in 1964.

With devaluation and the decrease in production costs, by 1966 petroleum company profits had once again reached extraordinary levels. AD was still in government, with Raúl Leoni as president, although the party no longer held a majority in the National Congress. The government then proposed via a Draft Bill to create a 'Selective Tax' that would only be applied to any net profits in excess of 15% of net fixed assets. The rate would be, at its highest, 38%, applicable when net profits exceed 50% of net fixed assets. The marginal government take, considering royalty, corporate income tax and the tax on extraordinary profits, would thereby have come to 73%. By leaving a minimum of 27% of any extraordinary income to the companies, the government sought to maintain the incentive for the companies to continue developing productivity (Pérez Alfonzo, 1967, p. 60).

The Selective Tax was conceived as an additional step, beyond the 50% Additional Tax created in 1948. Although the latter was not related to net fixed assets, the companies with the best lands would be the ones most affected by it in any event, and they were the ones that enjoyed the highest profit rates. But the major international oil companies were solely interested in maintaining uniformly the 50:50 arrangements in all the petroleum exporting countries.

In contrast, in their *Dissenting Vote* on the Hydrocarbons Law of 1943, Betancourt and Pérez Alfonzo (who cast the Vote), characterized oil production as a public service and contended, therefore, that the oil companies were entitled only to a customary profit and not to extraordinary profits. In 1948, their understanding had been that the Additional Tax was a step in this direction, and now, with the Selective Tax, they once again put forward the position they had held since 1943. But this time, the oil companies fought back.

In the USA, the companies paid, first of all, a proprietorial rent and, secondly (in certain cases) a differential rent, depending on the moment when the lease in question had been acquired. Subsequently, the lessee companies appropriated any other differential rent that might arise. Nor was the general fiscal regime geared towards taxing such rents, but the Selective Tax was meant for precisely this purpose.

Due to the determined opposition of the oil companies, the government failed to reach the legislative majority needed for the draft Bill to pass. Agreement was reached only after negotiation with the companies and the scrapping of the Selective Tax in a new Income Tax Law. In exchange, the minimum income tax rate was increased from 47.5% to 52%. Note that government's take of surplus thereby went, as a minimum, to 60% (=1/6% + (5/6)*52%). Profit sharing in a proportion of 60:40 had already figured in the discussions around the 1943 Petroleum Reform, and even the reform of the Income Tax Law in 1946. Moreover, reference to such a split had also been made in the Gentlemen's Agreement of 1959, where it was stated that the governments involved 'should be guided by a 60:40 formula as a minimum to bring themselves into line with the recent position assumed by Venezuela'.

In any event, the 52% rate reflected the prevailing corporate tax rate in the USA for normal industrial and commercial activities, including downstream petroleum activities; that is to say, refining, transportation and marketing.

However, the new Law negotiated with the oil companies also provided that:

> The Executive [...] may conclude a prior agreement with the taxpayer for a period of up to five years establishing the bases in accordance to which the values of [...] exported commodities at the Venezuelan loading port shall be determined. [...] The ex officio assessment shall not be applicable, in terms of prices, when an agreement such as that provided for in this paragraph has been entered into (Art. 41).

Hence, the National Congress ended up by delegating its sovereign tax powers to the Executive, thereby accepting a negotiated arrangement of fiscal reference prices, in accordance with the arrangement imposed by the major international oil companies with Saudi Arabia, Iran, Kuwait and Libya. In Venezuela, these fiscal reference prices were to be called Fiscal Export Values (*Valores Fiscales de Exportación* – VFE).

4. Corporación Venezolana del Petróleo

Betancourt and Pérez Alfonzo, already in 1947, had come up with a policy of *No More Concessions*. They believed that, after the massive grant of concessions in 1943 and 1944, whatever prospective lands remained ought to be developed by a national petroleum company, in cooperation with the private sector (Betancourt, 1956, p. 305). Notwithstanding, in 1956, Rómulo Betancourt also argued that:

> [...] the ultimate aspiration of all Venezuelan patriots would be the nationalization of the industry (Betancourt, 1956, p. 717).

And, more specifically still,

> [...] the most favorable solution would be the nationalization of the industry. That the state assumes the control of production, having previously indemnified the concessionaires, and thereby freeing the country from the presence in its soil of aggressive foreign economic and political forces (Betancourt, 1956, p. 740).

Now, in 1960, the Betancourt government created the first Venezuelan state oil company, the *Corporación Venezolana del Petróleo* (CVP). This company was authorized to engage in all the activities of the petroleum industry, from exploration and production to refining and domestic market supply, all the way to international marketing. The CVP was organized as an autonomous institute, under the tutelage of the Ministry of Mines and Hydrocarbons, and its activity was to be based not upon concessions but, instead, upon assignations of acreage with an indeterminate duration. The CVP, in turn, would be authorized to enter into "service contracts" with private capital for the areas assigned to it, but these contracts were to be definitively subject to the concessionary regime.

In 1967, in order to provide the space for the CVP to develop, the scope of article 3 of the 1943 Hydrocarbons Law – which authorized the National Executive to grant concessions – was widened. The article now provided for the possibility that the CVP would be able to pursue "agreements" or "mixed enterprises" with private capital, "as long as the terms and conditions set forth in any contract entered into pursuant to this article would be more favorable for the Nation than those set forth for concessions under the same Law". The Exposition of Motives referred to these "agreements" or "mixed enterprises" in general under the designation of "Service Contracts", which were conceived as a kind of second floor of the fiscal edifice, with the concessionary regime proper constituting the ground floor.

Just the same as concessions, these Service Contracts would also be subject to the tax sovereignty of the Venezuelan legislature and to the exclusive national jurisdiction. Concerning the latter, the following clause was to be obligatorily incorporated in all Service Contracts:

> "Any doubts and controversies of whatever nature that may ensue because of *this agreement* and which cannot be amicably settled, shall be decided upon by the competent Courts of Venezuela, and in accordance with its laws, and for no reason nor for any cause shall they give rise to foreign claims" (1967, Hydrocarbons Law, art. 3; quotation marks in the original; emphasize ours).

An identical clause had already been included in article 4 of the 1943 Hydrocarbons Law, albeit in that case reference was made to 'this concession'. Hence, the way was blocked for any attempt to limit, by indirect means (i.e. the contractual relationship with a state company), the sovereign rights of the state. Furthermore, regarding the issue of reversion:

> In the agreements, it shall be set forth that the lands and permanent fixtures, including the facilities, accessories and equipment that form an integral part of them, and whatever other goods acquired to fulfill the object of the agreements, regardless of the manner in which title thereto was obtained, shall be conserved in order to be turned over as a property of the nation, upon the extinction, for whatever cause, of the respective agreements.

Thus, when the terms of the agreements would expire, the CVP would have the appropriate infrastructure to ensure the continuity of productive activities. Moreover:

> The Chambers in joint session, duly informed by the National Executive of all pertinent circumstances, shall approve the contractual bases, taking into consideration the conditions established by them.

Article 3 of the Law also defined certain minimal conditions for agreements. For example, they were to have a maximal duration of thirty years and a specific area had to be defined in the contractual bases. In the case of exploration-exploitation agreements, once an area had been explored, its extension had to be reduced by 80%, by means of a sequential selection process.

Finally, the completed agreements had to be published in their entirety in the Venezuelan Official Gazette, but *not* approved by the National Congress. Thus, they would be open to critical review by any interested party and, if necessary, would also be open to being challenged in the national Courts on the grounds of non-conformity with the bases for contracting approved by the National Congress.

4.1. Service Contracts

In August 1970, the National Congress approved the contractual bases which would govern the service contracts for five exploration-exploitation blocks (GOV, 09/09/1970). The approved bases for all the blocks were essentially identical. In what follows, we shall make reference to the block that ended up in the hands of Royal Dutch-Shell.

In this case, the following was specified as the "Object of the Contract":

> The Contractor shall carry out with its own means and for its exclusive account and risk, albeit in the name and on behalf of CVP, all activities related with the exploration of the contract area, and the development and production of petroleum in the selected area [...].

Furthermore:

> For the elaboration, execution and supervision of plans, programs and budgets required for development, CVP shall have a broad operational participation, which it shall exercise through Committees and Sub-Committees which shall be constituted by equal numbers of representatives of CVP and the Contractor.

Thus, the CVP was being prepared to assume the productive activities in question, once the agreement would come to an end. For this reason, it was also stipulated that there be an obligation that:

> The Contractor's personnel shall be Venezuelan. Only in cases, when specialist technical personnel are required and not available in Venezuela, may [...] foreigners be hired for a limited time and on condition that they simultaneously train Venezuelans in the corresponding specialist task.

The CVP was also tasked with exercising, in cooperation with the Ministry of Mines and Hydrocarbons, "due care, to ensure that the Contractor maintain, and give proper use to, the goods whose title is to pass to the Nation".

Finally, the Contractor was also obliged to "pay the National Treasury, in the name and for the account and order of CVP, all taxes, contributions and sundry fiscal obligations set forth in the Hydrocarbons Law".

The CVP itself would also enjoy a share of the net profits of the Contractor, after the payment of bonuses, rents and royalties and income tax, whenever the net profits per barrel exceeded a threshold of 0.18 US$. At per barrel profit levels beyond 0.50 US$, CVP's share would reach 55%. Thus, the terms of the Service Contracts were effectively those of the failed 1966 fiscal reform.

In July 1971, the CVP signed the five agreements in question: one with Royal Dutch-Shell, one with Mobil and three with Occidental. The CVP also had the option, in case of a commercial discovery being made, of acquiring a shareholding participation. In the case of the Shell contract, this could reach 20%, but it was limited to 10% in the case of the Mobil contract and 1% in the Occidental contracts. The Agreements were duly published in the Official Gazette in December of 1971.

5. Peak Production

In 1970, the prevailing idea concerning reversion in Venezuelan political circles centered on mixed enterprises, constituted with Venezuelan capital, both public and private, and possibly complemented with a minority participation by foreign petroleum companies (Egaña and Maza Zavala, 1971). Furthermore, national public opinion took it for granted that the CVP would play a central role in this process.

But in 1970 Venezuelan petroleum production reached its peak, as was true also of US-American petroleum production, and for the same reasons: the depletion of traditional production areas. In Venezuela, this development also took everybody by surprise, even though it was by no means unforeseeable. In fact, in 1962, Venezuelan geologist Aníbal R. Martínez, applying the method developed by King Hubbert, had reached the conclusion that peak production in Venezuela would occur during the early 1970s; subsequently, in 1971, he estimated that the peak would occur in 1973, at 3.8 million b/d (Martínez, 1963, p. 61ff; and Martínez, 1972, p. 87).

However, the Venezuelan political world was not interested in this kind of prediction – and neither were the international oil companies. The latter limited themselves to blaming reduction in output on the pending reversion of the most important concessions – scheduled to happen between 1983 and 1985 – and especially on the increase in income tax rates and the sovereign setting of fiscal reference prices by the Executive from December 1970 onwards.

At this point in time, President Rafael Caldera and his Christian-democrat party Copei[1] were governing. Caldera attributed the decrease in Venezuela's production, to reprisals adopted by the companies in response to his nationalist petroleum policy. His successor, President Carlos Andrés Pérez (from the AD party) would attribute the decline to a policy of conservation. At no point was the possibility ever raised in the discussion that the country's production from traditional areas might have reached its historical peak.

6. Nationalization: National Productive Activity and International Source of Ground Rent

To start from 1970, the international petroleum scene had already undergone a structural shift whereby petroleum as a source of international ground rent suddenly became preeminent. Nevertheless, in 1973, the whole of Venezuela – as well as OPEC, the United States, and the rest of the world – was taken by surprise by the spectacular developments in the world petroleum market. As a consequence, despite the drop in Venezuelan output, fiscal income would not decrease (as had always been prophesied by the concessionaires) but, on the contrary, would expand explosively. Over and above the Reversion Law, the expectations radically changed in favor of the landlord state. It was then that the companies decided to beat a retreat, albeit by going forward. Instead of passively awaiting the reversion of the concessions granted between 1943 and 1945, the companies would seek that reversion be extended to the concessions granted between 1956 and 1957 and even to the service contracts that had been entered into by the CVP a mere two years before, in 1971.

In 1973, national elections were scheduled to be held in Venezuela. Due to a pact among the major political forces, dating from 1958, the issue of petroleum policy did not figure in the election campaign. Anyway, the election manifestos of the main contenders for the presidency – the eventual winner Carlos Andrés Pérez for AD and Lorenzo Fernández for Copei – did not contain even a hint that the petroleum industry might be nationalized or that reversion might be brought forward in time.

Nonetheless, a mere two weeks after the elections had concluded with the victory of Pérez and AD, at an act commemorating the fifteenth anniversary of the establishment of Cordiplan (*Oficina de Coordinación y Planificación*) and in the presence of the still incumbent president Rafael Caldera, senator Enrique Tejera París – a member of the National Executive Committee of AD – proposed "to study the nationalization of the petroleum companies". President Caldera, taken aback by this proposal, responded with the following comments:

---

[1] *Comité de Organización Política Electoral Independiente*, founded by Caldera in 1946.

> I believe that so far no one had ever put forward such a proposal in Venezuela. There had been some talk about the nationalization of the industry on the basis of the national company […]. Maybe it is worthwhile to analyze this because I have to say that, in truth, it had never occurred to me that it might be possible or appropriate to nationalize Creole, a branch of the Standard Oil [of New Jersey], Shell de Venezuela, a branch of [Royal Dutch] Shell, but given the authority of the one who came up with this concept, I believe that it is imperative not to let these words stand on their own, without making them the obligatory object of analysis and reflection (*El Nacional*, Caracas, 28/12/1973).

Two days later, the presidents of Creole and Shell de Venezuela paid a year-end visit to President Caldera. Upon leaving the Presidential Palace, they stated that:

> The potential nationalization of the petroleum companies that operate in the country, suggested on Thursday […] by Senator Enrique Tejera Paris, "constitutes a real possibility and point of view which is worthy of study", said […] Dolph, president of Creole […] pointing out that the relationship between companies and governments is changing everywhere in the world, and that Venezuela cannot be expected to be an exception to this trend, which means that it is imperative to study carefully all possible alternatives (*El Nacional*, 29/12/1973).

In any event, according to Dolph, it was up to the government to decide which was the best solution, and he added that:

> […] we are sure that a new relationship can include us in a very important role (*El Nacional*, 29/12/1973).

The president of Shell de Venezuela made similar statements. Finally, when Carlos Andrés Pérez took office as president in February 1974, he announced the nationalization of the petroleum industry. In formal terms, this measure would be presented as the 'anticipated reversion'.

6.1. Reversion Commission

Carlos Andrés Pérez promptly announced the creation of a Reversion Commission, in which all of the political currents in the country were to participate. The Commission would be entrusted with drafting a Nationalization Bill. Hence, there began a cycle of political discussion which was kicked off in May 1974 by Pérez's speech upon the occasion of the swearing in of the Commission. In December 1974, the Commission delivered its Draft Bill to the National Executive, Carlos Andrés Pérez, who in turn revised and modified it before sending *his* Draft Bill to the National Congress in March 1975. There followed an intense parliamentary scrutiny of both versions of the Bill, in which former presidents Caldera and Betancourt figured prominently. The process was brought to an end with another speech by President Pérez, in August 1975, when he affixed his signature to the Organic Law Which Reserves to the State the Industry and Commerce of Hydrocarbons (*Ley Orgánica que Reserva al Estado la Industria y el Comercio de los Hidrocarburos*) to which we will refer to as the Nationalization Law.

6.2. Carlos Andrés Pérez

6.2.1. No More Concessions and Speech Setting Up the Reversion Commission

In the Constitution of 1961, the granting of new concessions had been subject to the authorization of the Legislative Power:

> In no case shall the granting of new hydrocarbon concessions proceed [...], without the authorization of the Chambers in joint session, duly informed by the National Executive as to all pertinent circumstances, and in compliance with the conditions they set [...] (Art. 126).

Nevertheless, in 1962, Betancourt had asserted that:

> The word 'concession' has an unmistakable colonialist tinge and, for at least as long as I am governing, no more petroleum concessions are to be granted in Venezuela. I had the opportunity to discuss this problem with the leaders of the most important companies operating in the country, among them the managers of Creole [...], of Shell, of Mobil, etc., and they admit and acknowledge the idea that the government of Venezuela, through its state enterprise, may conclude service contracts that would not amount to concessions (Betancourt, 1962, p. 87f).

However, President Carlos Andrés Pérez, in the speech he delivered in May 1974 to mark the start of activities of the Reversion Commission, stated:

> Today, there comes to an end the turbulent and long historical process marked by shameful weaknesses, errors, equivocations, national humiliations and valuable experiences. It has been a road full of setbacks and frustrations [...].
>
> Today's decision is the fortunate, happy and triumphal synthesis of three decades during which the nationalist purpose was summarized and consolidated as a national determination, which found its first daring expression in the *Dissenting Vote* of 1943 [...]; and was asserted years later through the proclamation of the *no more concessions* policy denouncing them as the manifestation of colonialism. The idea of the rescue of petroleum was already alive then, and it stirs and becomes the conscience of the Nation [...] to erupt, defining paths and objectives, during the decade of the 1960s with the foundation of OPEC (Catalá, 1975, p. 15; emphasize ours).

In this way, Carlos Andrés Pérez perpetuated the historiography of Venezuelan petroleum devised by Betancourt and Pérez Alfonzo (and hence,

AD), and which divided the history of the country into two periods, the first one characterized by the selling out to foreign interests by oligarchic and dictatorial governments devoid of popular support, and the latter one by the nationalist positions assumed by democratic governments, that is to say, AD governments. In addition, he reasserted Betancourt's position from 1962, attributing a colonial tone to the word 'concession', and projected it backward into the past, presenting it as inherent to the *no more concessions* policy enunciated in 1947.

6.2.2. The Developed Capitalist Countries and the International Oil Companies

President Pérez had some very stern words for the developed capitalist countries:

> A new historical stage has opened up for Venezuela, Latin America and all other countries that are victims of the economic totalitarianism of the major developed countries (Catalá, 1975, p. 16).

He was referring to the way that these countries managed domestic production levels and import quotas for all kinds of primary products – mineral and agricultural – without taking into account the interests of exporting countries. In contrast, he took a considerably softer line as regards the international oil companies:

> It is no point to pass judgment or appraisals on the transnational enterprises that up to now have been managing the national petroleum business. Currently they are our partners and we should treat them as such. What we cannot nor should not admit hereafter is to transfer our transgressions, errors or omissions unto others, which ultimately only stand to benefit from our own irresponsibility, idleness or indolence (Catalá, 1975, p. 22f).

Hence, as the companies were 'currently our partners', it stood to reason that they were to form part of the new regime in one way or another. As a practical matter, in order to guarantee the continuity of petroleum activities, "there should at least be constituted four legal entities that will succeed Creole, Shell, Mene Grande and one for all the rest" (Catalá, 1975, p. 22). The CVP was to be included among "all the rest". Furthermore:

> At the helm of these new Enterprises must be left those Venezuelans who are currently managing the international Enterprises, thereby ensuring that the state will take over the administration of the business without any setbacks (Catalá, 1975, p. 21).

This was imperative because:

> [...] a genuine responsibility awaits us to manage efficiently an industry which moves in accordance with parameters never handled by Venezuelans, with a

state structure that we acknowledge to be inefficient and cumbersome; and regrettably almost devoid of the public spirit awareness that is both necessary and conditioning for the assumption of the stewardship of the petroleum destiny of the Nation (Catalá, 1975, p. 21f).

That being the case, it therefore followed that:

> In order to lead these Venezuelan Enterprises which are currently controlled from abroad by their parent companies, as far as their policy and fundamental decisions are concerned, there should be set up an organism or National Petroleum Enterprise, of the kind known in corporate law by the name of "Holding", which will be tasked with the highest responsibility for planning, guiding and supervising the four enterprises that shall be in charge of the daily operation of the Venezuelan petroleum industry (Catalá, 1975, p. 21).

Three national enterprises were to take the place of the three major international oil companies, in charge of production activities; and a forth national enterprise would cover all the other companies. Evidently, these national companies would retain their personnel in place, and also maintain the extant remuneration structure and human resources policies in general. In other words, their personnel would not become public employees. Furthermore, these entities would all be under the tutelage of a 'holding' company.

In contrast to the mandate for these companies, Carlos Andrés Pérez minimized the importance of the Ministry of Mines and Hydrocarbons, traditionally in charge of petroleum as a source of ground rent:

> [...] there does not exist within the Ministry any study that encompasses, and reaches conclusions about, the manifold technical, economic and legal aspects involved in bringing forward the reversion process (Catalá, 1975, p. 18f).

Furthermore, the Reversion Commission, which was being set up with much fanfare, would only be entrusted with the "determination of the formula for the calculation of indemnity", and with "setting the rules which guarantee the welfare benefits [...] of the workers" (Catalá, 1975, p. 19), but not with setting the guidelines for petroleum policy.

6.3. Report and Draft Nationalization Bill of the Commission

The Commission did not, as a matter of fact, limit itself to its assigned mission. In the Exposition of Motives of its Draft Bill, it asserted on the one hand that, "in order to manage the nationalized petroleum industry and bring it in line with the highest interests of the country, it is necessary to create a National Petroleum Administration" (Catalá, 1975, p. 53), and it duly provided that such administration would have to be created by means of a special law (at. 26). On the other hand, article 5 of the Draft Bill provided that reserved activities were to be carried out by "entities wholly owned by the state and created by means of special legislation". In its Report, the Commission specified that "any possibility to create mixed or capital participation enterprises for carrying out the reserved activities was totally rejected" (Catalá, 1975, p. 196). Petroleum, as a productive activity, would therefore be fully nationalized.

Furthermore, the Commission recommended that:

> The CVP be kept operating in its assigned areas and carrying out its current activities, until such time that the National Petroleum Administration would be established and take part in the general re-structuring of the industry, which will affect all operating companies belonging to the National Petroleum Administration. In this restructuring, it is recommended that much attention be given to the role that the CVP has played as a pioneering enterprise in the petroleum industry activity of the Venezuelan state (Catalá, 1975, p. 58).

Finally, the Report stressed that:

> The Ministry of Energy and Mines should be kept as the state organ in charge of formulating petroleum policy and as the mechanism for the normative control of the activities of that industry, with its operation readjusted to take into account the new perspective that arises from the exclusive activity of the state in the petroleum sector (Catalá, 1975, p. 59).

The Ministry of Energy and Mines would continue to be in charge of petroleum as a source of an international ground rent but, in addition, a National Petroleum Administration would be created and specifically entrusted with the supervision of the nationalized productive activity.

Moreover, the Nationalization Law, as indicated by its official title – *Organic* Law Which Reserves to the State the Industry and Commerce of Hydrocarbons –, was an organic law:

> [...] which means that, due to its hierarchical placement between the constitution and ordinary laws, the latter, in so far as they deal with hydrocarbons, would be governed by its provisions (Catalá, 1975, p. 63).

Indeed, the 1967 Hydrocarbons Law had not been an organic law. The Nationalization Law draft bill specifically confirmed that it was to remain in force. However, its last article provided that any previous legal dispositions not in agreement with its contents would thereby be repealed.

6.4. Draft Nationalization Bill of the National Executive

The National Executive received the Commission's Draft Bill, together with its Report, in December 1974. But it was only in March 1975 that the Executive sent to the National Congress what would be *its own* Draft Bill. *The concessionary regime would be abolished on December 31st, 1975* (art. 1). Art. 6 provided that the National Executive "shall create, with the legal configurations that it considers convenient, the enterprises that, in its judgment, are required" (Catalá, 1975, p. 128), and "one of the enterprises shall be entrusted the function of coordination, supervision and control of all the others, it being possible to assign to it the ownership of the shares of any of those enterprises" (Catalá, 1975, p. 144). The CVP was to be converted, from an autonomous institute under the tutelage of the Ministry of Mines and Hydrocarbons, into a mercantile society. In addition, and without any explanation whatsoever, the article dealing with the establishment of a National Petroleum Administration was eliminated. With the nationalization of petroleum as a national productive activity there would definitively disappear any consideration of petroleum as a source of international ground rent.

The Executive's Draft Bill therefore reflected the thought implicit in modern Economic Sciences. Accordingly, producers would optimize the extraordinary economic surplus, which was to be divided between an extraordinary profit, on one hand, and a ground rent for the landlord, on the other. However, in a situation where the landlord was also the producer, that person or corporation would enjoy the benefit of the whole of the extraordinary surplus, either as an extraordinary profit or as a ground rent.

Moreover, the Executive introduced a second paragraph in article 5, which was to open a space for private investment by means of association agreements:

> Article 5. The state shall exercise the activities set forth in Article 1 hereof directly through the National Executive or by means of agencies of its own. It may also sign any operative agreements that may be necessary for a sound performance of its functions. By no means, however, shall such negotiations affect the substantial nature of the activities hereby attributed.
>
> In special cases, and whenever it so suits the public interest, the National Executive or the said agencies, in exercising any of the activities herein referred to, may enter into association agreements with private entities; *maintaining, however, such a participation therein that will guarantee control by the state*, and establishing a fixed duration for such agreements. Authorization from both Chambers of Congress meeting in joint session shall be required prior to signing any such agreements, under such terms as may be set therefor, and upon the Chambers having been duly informed by the National Executive with respect to all matters pertaining thereto (Catalá, 1975, p. 103; emphasize ours).

No one questioned the first paragraph, with its reference to 'operative agreements', which already figured in the Draft Bill prepared by the Commission. Let us now consider the interventions in the debate on both Draft Bill in the Senate by former presidents Rafael Caldera and Rómulo Betancourt.

6.5. Rafael Caldera

In 1968, Caldera, Copei's foremost leader, had won the presidential elections. In 1975, the former President delivered a speech in which he would place the nationalization of the petroleum industry in the context of his historic vision of Venezuela as an oil country, and he would also highlight the benefits it would entail for the country.

6.5.1. Concessionary Regime

Caldera stressed how Venezuela had always been the target of a systematic campaign of disinformation on the part of the concessionaires:

> One of the mistakes that our generation has made with regard to the petroleum issue is to consider petroleum as something transitory and ephemeral [...].
>
> The petroleum companies came here with the idea [...] to obtain the greatest profit in the least time. [...] at each of the stages that we have lived through throughout their presence in the Venezuelan economic scene, [...] they led Venezuelans to believe the idea that petroleum would be running out very soon (Catalá, 1975, p. 171ff).

Hence:

> [...] we had to overcome threats, fears and rumors. Every time that the state had decided to take a step forward, ominous voices were raised and propaganda skillfully orchestrated, handled with admirable resources and technique, in order to convince us that we were heading towards failure, that we would kill the hen that lay the golden eggs (Catalá, 1975, p. 182).

Caldera enumerated some historical examples:

> Years ago the specter was Texas: the independent producers from Texas were the obstacle to our finding markets in the [USA] and only the influence of the top leaders of the major transnationals on the Washington government could guarantee us that [our barrels] would reach the American market, notwithstanding the vigorous and increasing opposition of the Texas producers.
>
> Then it was nuclear energy. According to the doctrine disseminated and embraced by all the spokespeople of the petroleum companies, not a single barrel of oil would have been burned anywhere in the world for some time now, everything would have been powered by nuclear energy. [...] The age of petroleum would not even extend to the physical depletion of the fields, as it was going to be substituted very soon by nuclear energy sources.
>
> They also talked about the Canadian pipeline that would make us lose that market [...].

They talked about Alaskan petroleum […], and about the sulfur which would close all markets for Venezuelan petroleum on grounds of the defense against environmental pollution.

And about Libya where every day the production indices rose […]. It appeared as if, in Libya, they would indefinitely have a 'bogeyman' to set up against Venezuela and, in fact, to enhance the psychological impact, technicians and machinery was transferred in the view and the presence of Venezuelans, so that they would become convinced that the petroleum business was coming to an end here because it was moving over to that country.

[…] and there was talk of five hundred thousand ton supertankers that would eliminate distances and thereby dispossess us of the advantages of some nearby markets (Catalá, 1975, p. 191f).

When Caldera took over the presidency in 1969, the international petroleum companies made a presentation to him about the possibility of grave threats for Venezuelan petroleum markets. At the time it was about Libya, Alaska and the re-opening of the Suez Canal. Given this, as far as Caldera was concerned, the fall in output from 1970 onwards could have no explanation other than as this being a further threat:

[…] when we applied a nationalist policy […] you know what the result was? […] from 1970 to 1972, oil production decreased by half a million barrels per day. They were trying to make us surrender through hunger. And if […] OPEC had not been successful, they would have allowed themselves the luxury of saying that […] the {Petroleum} Minister and I were a couple of madmen, who had led Venezuela to failure (Catalá, 1975, p. 200).

Caldera was at one with President Pérez in thinking that the issue of the price of petroleum was "one among those which most deeply wounds Venezuelan sentiment when it is examined" (Catalá, 1975, p. 276). Caldera did not doubt that when consumers paid fourteen dollars for a barrel of oil that they could previously buy for just two, they were paying a price that was more just. Reviewing the evolution of Venezuela's petroleum fiscal income, Caldera considered that "the petroleum scandal is not the price that we are now getting but the spoliation that we were subject to for so many years", and he made reference to 1935 – the year when Juan Vicente Gómez passed away – when "Venezuela received six US-cents for every barrel of petroleum exported" (Catalá, 1975, p. 175f).

However, as regards the development of the country's petroleum policy, Caldera took a very different stance to that of the AD leadership:

The collection of taxes kept improving through a sustained action in which we Venezuelans had […] a sense of […] historical continuity; maybe it is the only issue in which, despite the changes in governments and regimes, and the way in which mutations have arisen as a fruit of violence and spirits becoming

> possessed by a guerrilla mentality, every step of the way has been a rung on an ascending path (Catalá, 1975, p. 176).

In particular, making reference to the 1943 Petroleum reform, Caldera pointed out:

> [...] what in my opinion was most positive about it was [...] that it opened the way for the new legislation on Income Tax to be applied to the petroleum companies (Catalá, 1975, p. 179).

He likewise highlighted the importance of the "first mission of rapprochement" that Egaña had sent to the Middle East in 1949, "thus setting up the first pathway of contact for what was later to bear fruit with OPEC". (Catalá, 1975, p. 179).

But this historical continuity of national petroleum policy had always been met with the hostility of the international petroleum companies:

> [...] as far as I am aware, never have the major international oil companies supported Venezuela's steps in aspiring to a fuller sovereignty and a better deal for its petroleum; rather, I have the very reasoned impression that every time that they have been in a position to put obstacles in the way, they have done so, and far from being good fellow travelers or good partners – as has been said – trying to do their best for us, they have always tried to set up obstacles [...] (Catalá, 1975, p. 198).

Caldera was making reference, more specifically, to the parent companies and their representatives in Venezuela; that is to say, the upper executive echelons which confronted the Venezuelan state over petroleum as a source of international ground rent. He expressed his conviction that the problems that had arisen in the past regarding Venezuela's relationship with the USA were due to the powerful US-American international oil companies:

> The problem lies in that the North American government does not always make a distinction between the true interests of the people of the United States and the interests of major transnational companies which can be the enemies of that same people (Catalá, 1975, p. 212).

Finally, he expressed his hope that:

> [...] new and fruitful state-to-states relationships will be generated. A large part of the setbacks that we have suffered in our petroleum activity have been due to the fact that the major consuming countries, especially for us, the United States, have always resisted assuming the role that they ought to play. But if up until now they have left that role to be played by private entities, day by day they will feel more inclined to assume their full responsibility when they see that sovereign states, as opposed to transnational entities, are now playing the role of suppliers (Catalá, 1975, p. 217f).

Caldera was sure that, with nationalization:

> The country at large and the political country are going to get used to listening to, and respecting, the judgment of technicians and managers. It is in this way that the transformation of Venezuela's structures into those of a truly modern state will take place (Catalá, 1975, p. 215).

The nationalization of the industry was thus presented as definitely bringing to an end an inherently conflictive history.

6.5.2. Orinoco Oil Belt

The international petroleum companies had discovered the Orinoco Oil Belt in the 1920s. However, they were soon to conclude that the quality of the crude that they found, designated as "bituminous", made its exploitation non-viable with the technology then at hand. Thus, it came to be known as the Orinoco Tar Belt.

However, by 1971, the Service Contract with Mobil already provided for, as a "special advantage", a research program for the "Eastern Bituminous Tar Belt", valued at 800 thousand dollars:

> The project will consist of field experiments [...] for the recovery of crudes from the Bituminous Belt and shall be oriented to assessing the operational requirements, production rates and the recovery efficiency of the petroleum that might be obtained (GOV, 09/09/1970).

Moreover, the government then renamed the Belt as the Orinoco *Oil* Belt, on the basis that what it contained was extra-heavy oil, i.e. a fluid, as opposed to a tar-like natural bitumen. Even though it was estimated that the recovery factor would be no greater than two to three percent, it was expected that, with the progress of technology, recovery rates would before long reach ten percent. In view of its colossal magnitude:

> [...] the Orinoco Oil Belt completely changes the outlook for Venezuela in terms of its development plans. To anticipate the national planning as if petroleum were to run out in ten, fifteen or twenty years, is a completely different thing from planning the national development in the knowledge of what the Orinoco Oil Belt is and what it represents. [...] And today the thought arises of whether it might not be the most important petroleum reservoir in the world (Catalá, 1975, p. 204).

There could be no doubt that:

> [...] the eyes of the great powers are set on the Orinoco Oil Belt and it is there that the country has to have the strength of character of Simón Bolívar (Catalá, 1975, p. 205).

In particular, Caldera stressed:

> I am sure […] that all of the capital we might need, that all of the technology required, we shall have, because the Belt cannot be replaced with anything similar, it is a unique accumulation of riches and the means to exploit it will not be denied to the one who possesses it (Catalá, 1975, p. 206).

He concluded thus:

> The Orinoco Belt shall open the door to a fertile maternity, but with Venezuelan parents (Catalá, 1975, p. 207).

6.5.3. National Petroleum Companies

With the Executive's Draft Bill, the National Congress would have no function besides approving the nationalization as such. Petroleum policy would materialize setting up these national operating enterprises and their coordination by a parent company, which would be under the exclusive competence of the Executive. Caldera expressed his objections:

> I believe that the thesis of the Reversion Commission […] is correct. […] in order to create this group of enterprises, to set up a simple delegation of faculties in favor of the Executive is, in my opinion, not convenient (Catalá, 1975, p. 210).

6.5.4. Association Agreements

Caldera also voiced his opposition to the prospect of 'association agreements'. In specific reference to mixed enterprises, he asserted that

> […] the mere possibility of setting them up as is foreseen in the text of the law will not help, but only hinder. We who have had to fight with the transnational enterprises know that any chink that is left open shall be taken advantage of, bringing to bear all of their resources and power, in order to turn it into an open door. For the Government it would be better to negotiate with the transnational companies saying, I cannot set up mixed enterprises because Congress will not allow it. This is not an argument that will be available if the proposed official drafting is approved (Catalá, 1975, p. 195).

Caldera also did not believe that there would be any "major difficulty to find American technicians, principally, but also from other countries, to come and advise us and help us and solve for us the technological problems" (Catalá, 1975, p. 203), without any need to resort to association agreements.

6.5.5. Petroleum and Development

Caldera brought his intervention to a close by speaking on the topic of "Petroleum Resources and National Development", and his was the only speech that made reference to this subject. Caldera observed that the last budget of his presidential period, for 1973, had amounted to 14 billion Bs, whereas the first budget of the succeeding presidential period, that of 1974, exceeded 40 billion Bs. Nevertheless:

> [...] it would be a perilous error to say that we have too much money. What we have managed to get pales before the social needs that are indispensable to address. The fictional journalism in the major developed countries tends to create the impression that it is absurd that petroleum exporting countries have available to them quantities of money which are described as an irresistible tidal wave. Forty thousand million bolivar is a lot of money in relation to what we had before, and they allow us to expect that an administration conscious of the country's needs shall be able to overcome the negative aspects that still separate us from the developed world and keep us anchored to the rungs of underdevelopment, but it is not too much money as such. In truth, and only if they are harnessed efficiently, they might barely serve to cover the urgent and acute requirements of our social reality (Catalá, 1975, p. 218f).

Then, he drew a comparison between developed and underdeveloped countries and their requirements in terms of fiscal income:

> [...] our petroleum income, which feeds the bulk of Venezuela's fiscal coffers, quite apart from the urgent expenditures, involved in transforming the population, providing it with complete systems of education, health, accommodation, services and other fundamental aspects, has also to devote a considerable percentage to the construction of new petrochemical and steel complexes, to the exploration and development of the petroleum industry itself, to the construction and acquisition of naval units which will allow us to have an efficient fleet, to undertake all the infrastructure works that are being requested by the villages: in a word, to execute an effective economic and social development program. Far greater is the sum of resources that underdeveloped countries need to attain development, than what a developed country needs to continue attending to the needs of its people at a normal rhythm (Catalá, 1975, p. 219f).

6.6. Rómulo Betancourt

Rómulo Betancourt's speech was devoted, in the first place, to reasserting his version of the political historiography of Venezuela:

> [...] when talking about the transnationals, which have extracted from our country millions of barrels and billions of dollars [...] in excess profits, I shall not use any angry language but neither shall I use complacent language. And that is because I believe that if these companies have been able to plunder our country, it is because they counted on the complicity of rulers either without integrity or venal, who did not know how, or did not want, to defend the interests of Venezuela (Catalá, 1975, p. 242).

Betancourt took to the podium two days after Caldera, and he made it a point to deny the historical continuity that the latter had referred to. In particular, he asserted that:

> We also had the intention to send a delegation to visit the countries of the Middle East. This project could not crystallize because […] the government of […] Gallegos […] was toppled (Catalá, 1975, p. 253).

Regarding the regime which emerged from the coup of November 1948, he contemptuously observed:

> This regime barely fulfilled, transitorily, the idea of sending a delegation to the Middle East. It undertook a trip to various countries and returned to Venezuela. With neither shame nor glory, with more shame than glory (Catalá, 1975, p. 254).

Making reference to the Draft Bill of the Executive, he observed that he supported:

> […] in full that Article 5 which […] provides for […] the possibility of association […] contracts, which the Executive will not be able to carry out […] without the support of Congress, meeting in joint session of both Chambers. […] The circumstance may arise where an association agreement might be favorable and necessary for the interests of the country. That such an agreement could mean a new era of submission is something that I cannot conceive, because I have faith in Venezuela […], because I know that here […] there will be no more dictatorships, and only dictators are capable, whether out of venality or other causes, to fail to respect the national interest. I am sure that, starting with Acción Democrática and with the support of the other parties represented in Congress, any proposal which were contrary to the interests of Venezuela would be rejected (Catalá, 1975, p. 261).

6.7. Indemnity

The agreed indemnity for the nationalization came to one billion dollars, to be liquidated with a 480-million-dollar cash payment and the rest in installments paid over a number of years at an interest rate of 6%. The government negotiated at the same time a number of technical assistance contracts which would have a duration of three years, extendable. This assistance would be rewarded to the tune of 0.20 US$/barrel, net. These payments, together with the agreed indemnity, meant that the companies would continue to earn, for a few years more, essentially the same amounts that they had been making prior to the nationalization; that is, around 0.40 US$/barrel.

6.8. Signing of the Nationalization Law

The Executive's vision prevailed, thanks to AD's majority in the National Congress. There was only one modification introduced by the legislature which is worth highlighting: the operating companies would turn over to the parent holding company 10% of their net income from export sales, free and clear from income tax (art. 7 of the Nationalization Law).

When Carlos Andrés Pérez affixed his signature to the Nationalization Law, he made reference once again to "all those countries which are victims of the economic totalitarianism of the major industrialized nations" (Catalá, 1975, p. 267). Moreover:

> The economic order that made possible the exploitation of the natural resources of the poor countries is coming to an end. The peoples of the Third World are no longer willing to accept mean prices for their raw materials. A new international economic order has been proclaimed (Catalá, 1975, p. 282).

This new economic order was being promoted from the Organization of the United Nations. At its core was still the concept of *Commodity Agreements*, i.e. intergovernmental accords on basic products, referring to petroleum as much as other raw materials. However:

> In the meantime, the industrial countries are undergoing a crisis of misunderstandings which prevents them from accepting the fact that the exploited countries shall assume the active defense of their own interests and take their own decisions (Catalá, 1975, p. 282).

It appeared, then, that patience was what was called for, so that the industrial countries would be able to overcome their crisis of 'misunderstandings'.

6.9. Petróleos de Venezuela, S.A.

The government immediately created, in August 1975, Petróleos de Venezuela, S.A. (*Sociedad Anónima*: joint stock company), as a holding company, subject to private mercantile law (and, in effect, a private corporation) but with the state as a sole shareholder. In fact, this required that the legislation applicable to joint stock companies be reformed, as the latter required that such corporations have at least two shareholders. The national operating companies would also be constituted as joint stock companies, with PDVSA as sole shareholder.

With the disappearance of the concessionary regime, there arose a new regime, a *cessionary regime*, which was to be administered by PDVSA.

7. Conclusions

With the concessionary regime, the institutional seat of Venezuelan petroleum policy had been the Ministry. Since the First World War, such policy revolved around the maximization of the international value of the national natural resource. Later, the CVP, as an autonomous institute at the service and under the tutelage of the Ministry, benefited from an assignatory regime. But any agreement struck by the CVP with private parties would have as its foundation the concessionary regime, apart from requiring the approval of the contractual basis by both Chambers of Congress in joint session, as well as their publication in the Official Gazette. Finally, the definitive agreements also had to be made public, but *not* being approved.

But now, with the cessionary regime, PDVSA and its affiliates would become the institutional seat of petroleum policy, which would henceforth revolve solely around petroleum as a productive activity. *Any operative or association agreement with foreign private parties, governed by article 5 of the Nationalization Law, would no longer have the extinguished concessionary regime as a 'floor'.*

With petroleum incomes reaching hitherto unimaginable levels, the elimination of the concessionary regime did not elicit any criticism, neither in the entourage of Carlos Andrés Pérez – Betancourt and Pérez Alfonzo included – nor in the political world in general. Nobody suggested that it might be prudent to insist on conditions and rights which the nation had enjoyed on the basis of the concessionary regime, in the same way as had been done barely eight years before, in 1967, when article 3 of the 1943 Hydrocarbons Law had been reformed.

As far as the problems of the past were concerned, Carlos Andrés Pérez blamed the developed capitalist consumer countries, while Rafael Caldera blamed the major oil transnationals, but they agreed that nationalization would put an end to the abuses perpetrated by both. The speeches that these politicians delivered reflected a total lack of comprehension and, even more so, the absence of the very notion of how a capitalist system might relate to land tenure. It never even crossed their minds that, within such a system, developed countries would define and defend an international regime aimed at keeping international ground rent to the lowest possible levels, and that the petroleum multinationals would put it into practice. Any success on their part would be to the benefit of consumers, in much the same way as consumers would benefit from the advances in productivity: through competition.

With the events in the Middle East in 1973, it could no longer be doubted that the international oil companies were completely tangled up in the concessionary regime, and unable to continue fulfilling their function of keeping the landlord states in check. Their capitulation at that point was, in reality, no more than a retreat, meant to create some breathing space in order to develop a new international regime. However, the misgivings that were voiced in discussions about the Nationalization Law were that the association agreements could herald a return to the concessionary regime of the past. Such fears completely overlooked the fact that the essence and whole point of the nationalization, as carried out in practice, had been precisely to put an end to the concessionary regime and to put a different one in its place, one that would once again allow the international petroleum companies to fulfil their role of representing global consumers against the landlord states.

The new arrangements would see the replacement of the Ministry, a regulatory entity focused on petroleum as a source of international ground rent, with PDVSA, a novel regulator that would focus on petroleum as a productive activity. It would be up to PDVSA and its affiliates to defend their entrepreneurial liberties against the landlord state. The most effective argument that they would have at their disposal would be that any restriction to such liberties, which might have been justifiable against the foreign oil companies and in the context of the concessionary regime of the past, no longer made any sense if applied to PDVSA, a national state oil company.

However, before dealing with Venezuelan petroleum policy after nationalization, it is necessary to review the topic of *Petroleum and Development*.

Part V.

Venezuela: Rent Capitalism

In 1899 General Cipriano Castro seized power in Venezuela by force of arms, at a time when the country was in dire economic straits, which prompted him to suspend payments of the international debt that the country had been carrying since its Independence. He subsequently also suspended payments on the private debt contracted with European and US-American citizens who operated in the country, accusing them of financing military rebellions against him.

In response, in December 1902, German, British, and Italian warships shelled the main Venezuelan ports and (in their capacity as creditor countries) occupied the customs buildings in order to guarantee the payment of the debt. As for the private debts, they demanded that the Venezuelan government accept international arbitrations, which were to have their seat in The Netherlands.

These arbitrations only concluded after vice-president General Juan Vicente Gómez took advantage of Castro's temporary absence, in December 1908, to usurp the Presidency, with the explicit support of the USA. In the immediate aftermath, Gómez would agree to settle all pending disputes; later, he would pay all sums due. Gómez was to stay in power – even though he would not always occupy the Presidency of the Republic – for the following 27 years, until his death in December 1935.

1. Petroleum: National Productive Activity and Source of International Ground Rent

At the beginning of the 20th century, even though Venezuela was a peripheral capitalist society where the great majority lived in rural indolence, it nonetheless boasted a modern legislation, based on French legislation, which pointed the way towards capitalist development. In this corpus of legislation, minerals were considered as a free property of the state, which meant that any productive activity involving such minerals required the approval of the state; that is to say, a concession. Then, the international oil companies turned up, and they developed, in one of the most backward agricultural exporting countries in Latin America, a modern and export oriented petroleum sector (Baptista, 1984). In Venezuela, a sparsely populated country lacking even the most basic infrastructure, the development of the petroleum sector gave rise to multiplier effects of major significance. The companies had to create their own infrastructure, and they also had to resort to a more qualified foreign labor force, but also to a large unskilled labor force from the Caribbean. Let us clarify that, for practical reasons, the 'petroleum sector' – in Venezuela as in other oil countries – comprises, apart from exploration and production, the refining, transportation and marketing (national and international) of crude oil and petroleum products.

The first concessions, granted before the First World War, ended up in the hands of European companies. Already then, the Venezuelan government viewed petroleum optimistically, as a new and important source of fiscal income, although without expecting anything like what was to happen when exports began in 1917. Then, US-American companies arrived, offering bonuses, rents, and royalties, in accordance with US practices. With this, the government became conscious of the extraordinary potential of petroleum as a source of international ground rent. Graph V.1 contains the most basic data necessary to analyze petroleum both as a national productive activity and as a source of international ground rent[1].

Figure V.1

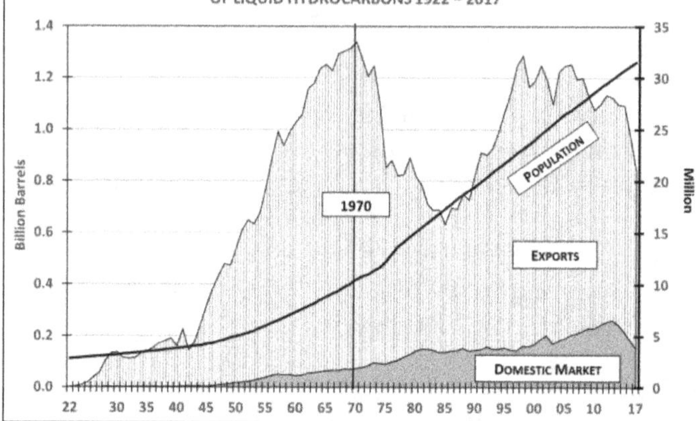

1.1. National Productive Activity

By 1929, Venezuela was already the most important petroleum exporting country in the world. The petroleum industry counted 27,221 workers, of whom around a third were foreigners. Subsequently, in 1948, with the boom of the second post-war period, the industry would reach the maximum employment level recorded during the concession era, at 61,077 workers, of whom around 10% were still foreigners. In 1970, when production reached its historical peak, the industry workforce was down to only 23,993 workers, of whom less than 5% were foreigners.

Despite the significant increase in the workforce and the net capital stock[2], after the nationalization of the industry, the output reached in 1970 would never be achieved again (Figure V.2). Figure V.3 shows the output per worker,

[1] The statistical information in this Part V are based, unless indicated otherwise, on Asdrúbal Baptista: *Bases Cuantitativas de la Economía Venezolana: 1830-2008*, Caracas 2011, as well as subsequent personal communications.

[2] For the concept of net capital stock, see Baptista 2011, Chap. IV. In essence, it reflects the real net fixed assets and not the net fixed assets in financial statements, for the purpose of income taxation.

as well as the net capital stock per barrel produced. The first indicator reached its maximum in 1970 and then decreased again, and in 2011 it reached a level last seen during the 1930s. The net capital stock necessary to produce one barrel of petroleum, in contrast, simply grew and grew.

Figure V.2

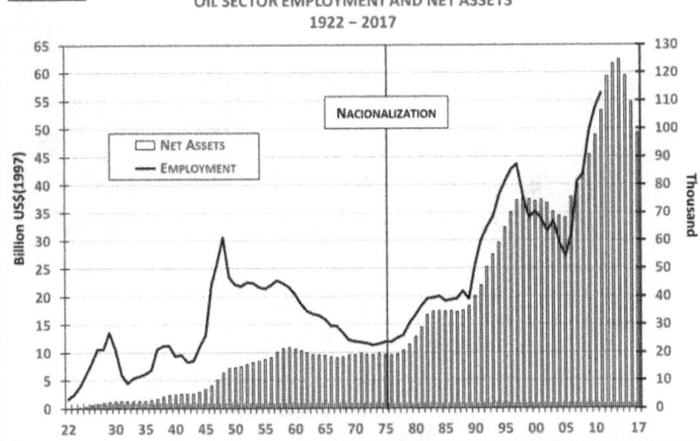

From the point of view of petroleum as a productive activity in the national context, the petroleum industry in 1930 accounted for 26% of the net non-residential capital stock in the country. With the second post-war boom, and the construction of the two major refineries located in the Paraguaná peninsula, in 1949 the industry came to represent 37% of the net non-residential capital stock in the country (Figure V.4). This percentage subsequently decreased but, after nationalization, it started going up again and, towards the end of the period under consideration (i.e. 2017), it came to around 17%. As for the employment levels in the industry, a similar pattern can be observed if this indicator is compared against the size of the population. In short, from the standpoint of both capital and labor, the petroleum sector was and continues to be extraordinarily important.

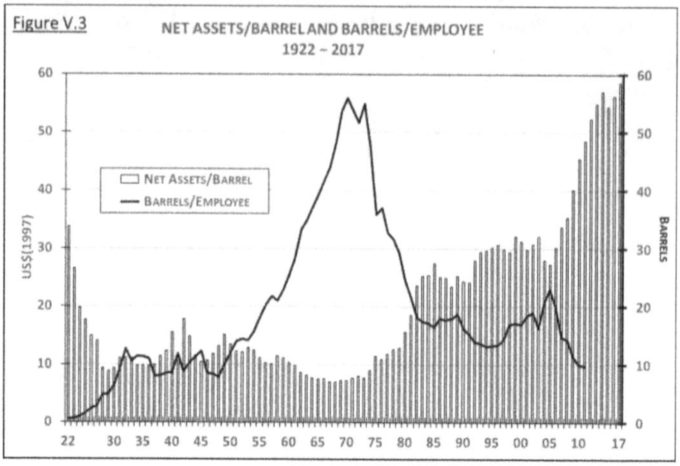

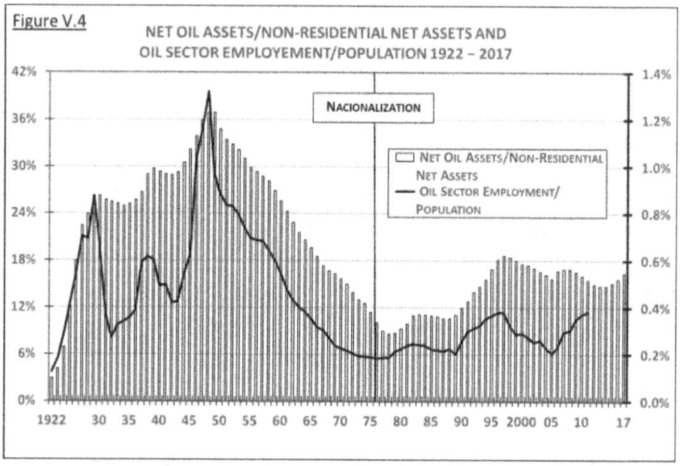

1.2. Source of International Ground Rent

The fall in Venezuelan petroleum exports, from 1970 onwards, coincided with an increase in the prices of petroleum in the international market and, therefore, in the price of the Venezuelan crudes and products export basket (Figure V.5).

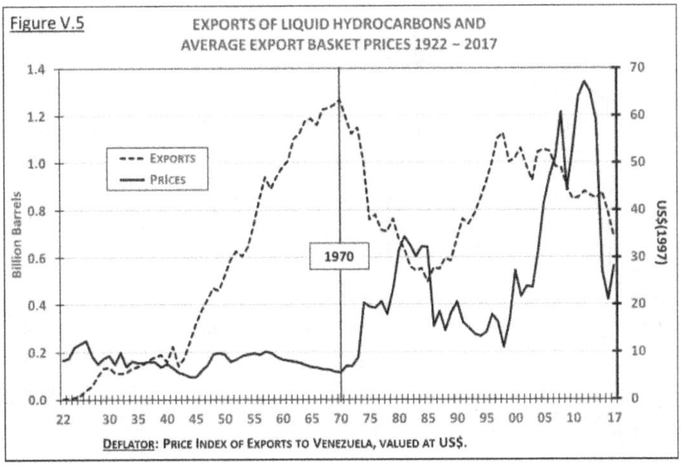

Figure V.5 EXPORTS OF LIQUID HYDROCARBONS AND AVERAGE EXPORT BASKET PRICES 1922 – 2017

In what follows, we consider as a net 'customary' profit, a 10% return on the net capital stock. Any returns beyond this 10% threshold are deemed to be extraordinary surpluses.

When exports began to decline in 1970, the rise in prices up to 1974 was such that it more than offset the shrinking volumes and the corresponding increase in production costs per barrel (Figure V.6). Furthermore, the trend for the extraordinary international surplus (EIS) per exported barrel continued to rise throughout the whole period, notwithstanding the ensuing tug of war between petroleum exporting and importing countries as regards prices. However, once the concessionary regime had been liquidated in 1975, prices were unstable. In 1998 and, once again, in 2016 and 2017, the EIS decreased to practically zero, a situation that had never arisen before.

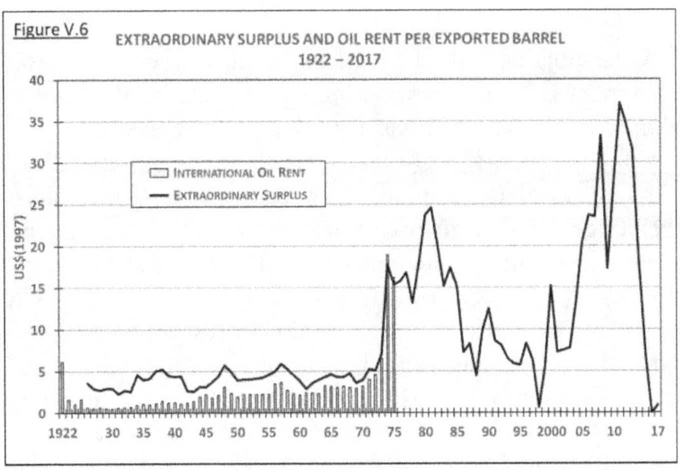

Figure V.6 EXTRAORDINARY SURPLUS AND OIL RENT PER EXPORTED BARREL 1922 – 2017

Furthermore, it is worth noting that the state, throughout the duration of the concessionary regime, benefited – including by way of bonuses – from a rising international petroleum rent (IPR). However, up to 1973, the concessionaires always appropriated a significant portion of these surpluses in the form of extraordinary profits. Regarding the distribution of these surpluses, subsequent to nationalization, we shall have more to say further on.

But even though petroleum as a source of international ground rent became increasingly important, exported barrels per capita reached their maximum already in 1957 and, in 2017, fell to the level of 1927, i.e. to the level they had attained ninety years before. Furthermore, given the continuous growth in population, the international petroleum rent per capita showed a downwards trend from 1974 onwards, albeit with an important increase between 2004 and 2014. (Figure V.7)

Figure V.7

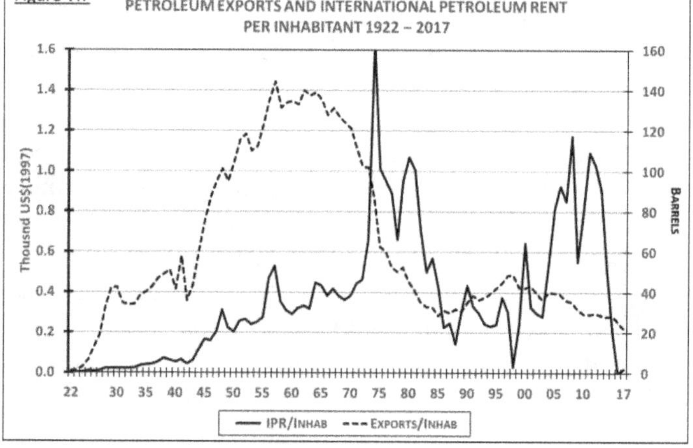

1.3. Conclusions

In 1917, Venezuela became a petroleum exporting country and, in 1928, it became the most important petroleum exporter just as the market boom that followed the First World War was reaching its conclusion.

During this first period, petroleum fiscal income amounted to 129 million Bs, while the petroleum capital stock came to 654 million Bs. But the multiplier effects of foreign investment were very important in relation to fiscal income. Even though the industry itself was exempt from import duties, its employees and service providers were not. In fact, the increase in the collection of import duties throughout the 1920s exceeded by a significant margin the fiscal income from petroleum as such (B. Mommer, 1990; Espinasa and B. Mommer, 1991). In other words, the first petroleum boom in Venezuela was a boom based on productive activity, whereas appropriation of international ground rent was still secondary in importance. In any event, the result was a fiscal bonanza that had no precedent in the country's history.

However, the boom that followed the Second World War was not only a boom of petroleum as a productive activity but also as a source of international ground rent. Even though the magnitude of the two subsequent booms for petroleum as a source of international ground rent – the first one between 1973 and 1985, and the second one between 2004 and 2014 – proved to be considerably greater in magnitude, they arose in the context of an industry in decline, with lower export levels and subject to political confrontations that we shall deal with later. With the end of the concessionary regime, Venezuela, as an oil country, had to face up to and deal with an unstable world market.

2. From one Venezuela to Another

At the beginnings of Venezuela's history as an oil country, the issue of ground rent – and even of an international ground rent – was obvious, just as it had been for the physiocrats and the classical political economists. But even though the starting point was the free property of the state over mineral deposits – in line with French practice – the reference soon became the private ownership of mineral deposits in the United States, which led to the corresponding demands for remuneration on the part of both the landlord state and private landowners (Baptista and B. Mommer, 1987).

2.1. Concession Dealers

The first concessions in Venezuela were granted to Venezuelan citizens who acted as middlemen on behalf of the international oil companies, and received commissions from the latter. But from 1917 onwards, once petroleum had become a source of international ground rent, there arose the question of who should be the beneficiary of such rent. The Minister for Development, Gumersindo Torres, considered that in lands under private ownership, the beneficiaries had to be the landlords, which is why the first Hydrocarbons Law of 1920 granted them the corresponding preferential rights. In fact, 2,300 landowners availed themselves of the opportunity to obtain concessions covering their lands, as a way of obtaining bonuses and, sometimes, also overriding royalties upon the transfer of these concessions to interested international oil companies. Subsequent hydrocarbons laws never again granted landlords such privileges, but the traffic in concessions continued.

Minister Torres' vision was one rooted on the reality of a primitive capitalist country which exported agricultural commodities and whose dominant class was landowners. Vicente Lecuna, president of the Bank of Venezuela (an entity that managed the finances of the government) [3],

[3] This was a private bank, not to be confused with the Venezuelan Central Bank, founded in 1940.

embraced an opposing vision. Lecuna did not doubt that public ownership was to be maintained, being *a historical legacy that could not be questioned*, as it had been confirmed by the Liberator himself. Furthermore, he anticipated that such ownership would provide the state with the fiscal resources necessary to modernize the country. Lecuna's vision would finally come to prevail in 1943 with the petroleum reform. The Exposition of Motives of the Hydrocarbons Law unequivocally affirmed that mineral resources were part of the public domain, and for good measure denounced:

> [...] the vacillations that appeared in some of our first mining laws, and tends to ignore the constitutional principle of uninterrupted tradition in the country, concerning the state's ownership of mines [...]; and that the surface owner, as such, does not have any rights whatsoever over mineral deposits.

The 1943 Hydrocarbons Law put an end, once and for all, to intermediaries. Concessions would thereafter be granted directly to the international companies, and any subsequent transfer would have to be authorized by the Ministry.

2.2. Decline of the Agro-Exporting Country

During the 1920s, the petroleum industry underwent an extraordinary boom, easily surpassing in importance, towards the end of the decade, all other traditional exports put together. The 1929 world economic crisis translated into a relatively minor and temporary decrease in petroleum exports. In fact, in 1930 – on occasion of the hundredth anniversary of the death of the Liberator – General Gómez took advantage of the fiscal bonanza in order to liquidate the whole of the external debt that the country had been carrying since Independence, as well as the bulk of private sector debt. Furthermore, he promoted the creation of national infrastructure, an indispensable requisite for any economic development and which the country was almost totally devoid of.

However, traditional agricultural exports – first and foremost, coffee – fell into the throes of a profound crisis, from which they were never to recover. The crisis in the international market only highlighted the dismal state in which Venezuelan agriculture found itself. In the words of Alberto Adriani:

> Our agriculture is rudimentary. There is not a single crop for which Venezuela could serve as an example to the world (Adriani, 1937, p. 247f).

Nevertheless, in his judgment:

> Agriculture and stock breeding are today and shall be tomorrow the principal bases of the prosperity and greatness of the country (Adriani, 1946, p. 191).

He attributed no importance whatsoever to petroleum:

> We should not be mistaken in assessing the changes that have come in the wake of the boom of the petroleum industry in Venezuela [...]. She is in foreign hands, and from an economic point of view, it is a foreign province enclaved in the national territory and exerts a relatively insignificant influence on the economic prosperity of our people (Adriani, 1937, p. 207).

For Adriani, extractive industries were:

> [...] precarious factors of prosperity. When the mines, whose main benefits shall have accrued to foreigners, become exhausted, the country shall have to shoulder the damages and pay the costs that the demobilization of those industries will imply (Adriani, 1937, p. 121).

Thus, Adriani envisaged a transitory and ephemeral mining boom, akin to the gold rushes of the USA. The oilfields could become depleted from one moment to the next. Hence, he concluded:

> *We should not insist on petroleum* (Adriani 1937, p. 129; emphasize in original).

Then, in 1934, when the USA devalued the dollar and the gold parity of the bolivar went from 5.20 Bs/US$ to 3.09 Bs/US$, the situation of the traditional agro-exporters became dramatic:

> Up until last year, the dollar in Venezuela kept a value which made the growing of coffee and cocoa mediocrely worthwhile [...]. With the newly devalued dollar or, what amounts to the same, the expensive bolivar, the prices of our export products have become risible, as they do not cover even the costs of processing the beans and are ruining all interested parties (Adriani 1937, p. 263).

Adriani's proposal was to devalue the bolivar by one hundred percent. But in this he was opposed by Vicente Lecuna, who argued that a devaluation would work in favor of the foreign oil companies, to the detriment of national interests. Indeed, the expenditures of these companies in the country were in bolivars – with the sole exception of percentage royalties, originally denominated in dollars – so the non-devaluation of the national currency would imply an increase in the companies' cost of production, in dollar terms. The costs of production themselves, such as salaries and wages, were generally in bolivars. Nevertheless, despite the significant reduction in the revenues of the oil companies in bolivars, due to the increase in the IPR, as well as the production costs (also in bolivars), the concessionaires were going to continue to benefit from extraordinary profits. Venezuelan petroleum would remain competitive.

In the short term, devaluation would have meant very significant losses in the petroleum sector, and any potential benefits accruing to traditional exports in the long term were very uncertain at best. Thus, the bolivar was not devalued and, on the contrary, the state began to *subsidize* traditional exports, *buying* the corresponding dollars at preferential exchange rates.

2.2.1. Arturo Uslar Pietri

The discussion around the potential bailout of the agro-exporting Venezuela did not end there, however, and Arturo Uslar Pietri became the most prominent advocate of this conception[4]. On July 14, 1936, he published an article on "Sowing the Oil" in the newspaper *Ahora*. He expressed anguish at the sight of the 'destructive economy' prevailing in Venezuela, 'which sacrifices the future for the present'. At that point, a third of the government's ordinary fiscal income came from petroleum, and another equally important part arose from the multiplier effects of the productive activity. Thus, a danger could already be glimpsed that the Venezuelan treasury could find itself in the extreme situation of:

> [...] paying the whole of the Budget solely with the income of mines, which can be simply translated in other terms thus: turn Venezuela into an unproductive and idle country, an immense parasite living off petroleum, swimming in a momentary and corrupting abundance and heading for an imminent and inevitable catastrophe (Uslar Pietri, 14/07/1936).

The catastrophe would be the exhaustion of petroleum, which Uslar Pietri saw as imminent. The abundance he saw as corrupting because it was not rooted in national productive activities. For the time being, the portion of the government budget dedicated to investment did not even come to one-seventh. However, it was:

> [...] necessary that the expenditures intended to create and guarantee the development [...] of a progressive economy reach at least a level of parity with the mining income (Uslar Pietri, 14/07/1936).

Uslar Pietri fully shared Adriani's opinions about the state of agriculture and animal husbandry in Venezuela, which continued to decline sharply. Nevertheless:

> The only wise and redeeming economic policy [...] is to transform the mining income into farming credit, stimulate a modern and scientific agriculture, import studs and fodders, replenish the forests, build all the reservoirs and channels necessary to normalize irrigation [...] mechanize and industrialize the countryside [...]. This would be the true action of national construction (Uslar Pietri, 14/07/1936).

---

[4] Alberto Adriani passed away suddenly in august 1936, at the age of 38.

According to Uslar Pietri, the current public expenditure would have had to be financed entirely with taxes from the non-oil sector. He then brought his article to an end with the slogan: *sowing the oil*.

For Uslar Pietri, the future of Venezuela would still be that of an exporter of agricultural commodities, even though his language would soon conform to that of the Economic Sciences. As the author of the first book on the Venezuelan economy, he would no longer speak about a 'destructive economy' but would instead assert that:

> Oil is neither a crop nor a rent, but the continuous consumption of a capital deposited by nature in the subsoil (Uslar Pietri, 1945, p. 243).

It was no longer a matter of *investing* the fiscal income of oil, but of *re-investing* it, otherwise the country would be decapitalized

2.3. Petroleum Exporting Country

2.3.1. Manuel R. Egaña

In 1938, General Eleazar López Contreras, President of the Republic, appointed Manuel R. Egaña as Minister of Development. Egaña then visited the oilfields and realized their impressive industrial structure. He immediately abandoned the vision he had thereto harbored of oil as an ephemeral bonanza, a vision he had shared with Adriani and Uslar Pietri. But now, for Egaña, there could be no doubt that the companies were in Venezuela to stay, following the pattern set by the coal industry in the USA and Europe.

Egaña henceforth turned the whole of his attention to the question of investing the IPR. He had no doubt that investing the IPR to develop infrastructure and to place the latter at the free disposal of the nation was a sound policy, being as, in his words: "Money, there is aplenty" (Egaña, 1941, p. LXIV). The scarce resource was human capital, which meant that public works had to be conceived bearing in mind its scarcity. For instance, whereas in the USA, the optimal solution for building a freeway might involve an asphalt-paved road that would thereafter receive the necessary upkeep, in Venezuela the optimal solution would likely be a road made from concrete, which would be more expensive up-front but would require much less maintenance. The workforce required for infrastructure projects would come from abandoning less productive activities in agriculture, on the one hand, and on the other from immigration, of European origin. Net migration was always a positive figure throughout the whole of the concessionary period, apart from some exceptional years. In real terms, all of this workforce was remunerated by *importing* whatever was necessary.

Egaña also suggested that:

> The state has to stimulate the reinvestment in the country of the profits and returns obtained by the foreign capitals, so that they can help with the economic development of Venezuela (Egaña, 1941, p. LIV).

As regards the nature of petroleum income, he made the following observation:

> The extraordinary return that the petroleum industry produces, does not constitute a new source of wealth which might be added to those that we already had, rather [...] it has come to substitute for those sources reducing some and sterilizing others. Now, when a country endowed with diverse sources of wealth sees the exploitation of one of them to expand uncontrollably, on account of circumstances at a particular juncture, it is under a strict duty to correct such deformation, fostering and strengthening the disturbed economic activities. The Venezuelan state cannot shirk the imperative of promoting and strengthening the national agricultural and manufacturing industries. Without the protection of the state, those industries would vanish from one year to the next, and this is not something that would be allowed by the two thirds of the Venezuelan population which makes its living off them.
>
> [...] The Venezuelan state has to carry out an intense economic policy on account of the fact that a considerable proportion of the national income ends up in its coffers. (Egaña, 1939, p. IX f)

The non-devaluation of the bolivar versus the dollar also meant a significant transfer of the international petroleum rent to the public which consumed imported goods, as imports became cheaper. Nevertheless:

> Import duties are [...] the most powerful instrument whose indispensable use maintains the rhythm of our economic activities, which we are not only obliged to maintain but are actually required to increase. The Venezuelan market should be reserved to what can be produced in Venezuela, provided that this does not entail sacrifices by the consuming public in the interest of particular persons or groups [...]. It is necessary to add, in view of Venezuela's peculiar circumstances, that one should not be put off by the prospect of increasing prices on account of the establishment or the increase in a tariff, if this is necessary to create a new industry, which is to provide employment for the many without giving rise to an exorbitant benefit for the few (Egaña, 1941, p. LXVIII f).

Ernesto Peltzer

Egaña was appointed Minister of Development with the specific mandate of setting up the Venezuelan Central Bank (BCV). In 1939, Ernesto Pelzer, a renowned German economist, arrived to Venezuela and joined the BCV at its foundation in 1940[5]. Peltzer provided a scientific foundation for Egaña's primarily empirical observations. In a 1944 publication, Pelzer drew a rigorous conceptual distinction between petroleum as a productive activity, on the one hand, and as a source of international ground rent, on the other:

> [...] incomes derived from exports broadly understood, [...] break down into [...] incomes arising directly from productive activities dedicated to exports as well as incomes which also arise from exports but entail no expenditure of productive forces, for instance, taxes and royalties on the exploitation and export of hydrocarbons (Peltzer, 1965, p. 150f).

Peltzer was especially well qualified to opine on the subject. Indeed, in the literature on international economics, the issue of unilateral transfers had been amply discussed after the First World War, on account of the reparation payments that defeated Germany had to make to the victorious Allies. Prominent personalities had participated in this debate, with the best known being John Maynard Keynes. The general conclusion of this debate was that, as a consequence of such transfers, price levels in the payer country decrease, while they go up in the recipient country.

That same year 1944, José Antonio Mayobre, in a thesis on The Parity of the Bolivar (*La Paridad del Bolívar*), supervised by Peltzer, asserted and verified that the "inflow of currency caused by petroleum exports can be likened to a unilateral transfer" (Mayobre, 1982, p. 373f). For his part, he then confirmed statistically that during the period 1920 to 1940, "the high cost of living in Venezuela, which constitutes a chronic phenomenon throughout the period we refer to, finds its ultimate explanation in unilateral payments" (Mayobre, 1982, p. 385).

2.3.2. Rómulo Betancourt

The position of AD with regard to the Bill for the 1943 Hydrocarbons Law, summed up in the *Dissenting Vote* cast by Pérez Alfonzo, was limited to advocating for the popular distribution of the IPR, for the benefit of a needy people. However, on the basis of the doctrine that the petroleum industry was a public service, the industry would only be entitled to a reasonable profit while all extraordinary surpluses would rightfully belong to the state.

After the October 1945 coup d'état, the Revolutionary Junta presided by Betancourt (and with Pérez Alfonzo as Minister of Development, responsible for petroleum related matters), implemented the popular distribution of petroleum fiscal income. Writing from exile in 1956, Betancourt noted in particular:

---

[5] Peltzer arrived to Venezuela as a refugee from Nazi Germany, as his wife was of Jewish ancestry.

> As the taxes levied on the most profitable economic activities could cover the larger percentage of public expenditures, it was suitable for the nation to reduce the burden of direct taxes on the income of employees, of medium and small businessmen, of the weakest strata, in sum, of the Venezuelan economy. This measure was complemented with the elimination of a significant number of indirect taxes (Betancourt, 1956, p. 244).

Furthermore:

> If we were confessed defenders of a general policy of high incomes for the working sectors, the state as an employer had to set the example, improving the wages and salaries of those who worked as employees or laborers at its service (Betancourt, 1956, p. 298).

The salary levels established by the state would also set the pattern for the private sector. Moreover, public spending increased very significantly in education and health, in the development of public services and infrastructure works, in the context of an extraordinarily rapid country-wide urbanization process. As a consequence, "the country, thanks to the salary increases and the considerable rise in the National Budget from 1946 onwards, became a market with a great purchasing power" (Betancourt, 1956, p. 381). The conditions had therefore been created for:

> [...] going to the very root of the economic and social problems of the nation [...]. It was necessary to transform the paltry output, limited to a few commodities, into a diversified and abundant production (Betancourt, 1956, p. 310).

In conclusion, what was required was:

> [...] *sowing the oil*. Thus, a transitory and perishable wealth would provide stable foundations to an ever-lasting Venezuela (Betancourt, 1956, p. 293; emphasize in original).

The slogan created by Uslar Pietri in 1936 had already been adopted nationally in its generic sense. However, even if "Venezuela had to strive to produce the major part of what it consumed, both in terms of the products of the land and of elaborated products" (Betancourt, 1956, p. 380), the idea certainly was not to give priority to agricultural and stock raising to rescue the agro-exporting Venezuela of yore:

> Our party, in its program [..] had raised and defended the thesis of the industrialization of Venezuela [...]. We had the advantage that petroleum income gave us the capability to accelerate an industrializing process, slower and more difficult in other under-developed countries that did not have such an exceptional source of income.

> [...] The competitive exportable agricultural products, typical of tropical zones – coffee, cocoa –, already subject to the competition of European colonies in various continents, offered unexciting prospects. Whereas the situation was very different regarding the prospects of manufactured goods which might be produced in the country, and which could find an outlet in the vast Latin American market. It was necessary, therefore, to give an impulse, in a planned manner and backed by a firm decision, to the industrial development of the nation (Betancourt, 1956, p. 380).

But being as the intention was to industrialize the country, and not simply to rescue the traditional agro-exporting sector, Betancourt was of the opinion that:

> [...] in Venezuela, the action of the state to stimulate and guide production is an imperative necessity. The financial might of the state, derived from the fact that it is the recipient of the bulk of the income provided by petroleum, renders the fulfilment of such an obligation both viable and unavoidable. The Venezuelan state cannot just sit back to wait for private enterprise to develop and diversify production [...]. It has to act [...] as a state which stimulates, finances and guides the economic activities which tend to make domestic production more abundant and varied; and, as an entrepreneur state, to develop certain activities directly linked to the public interest (iron and steel industry, electrification, radiotelegraph and telephone communications, transport) (Betancourt 1956, p. 315f).

Nevertheless, in 1948, with the dizzying increase in petroleum fiscal income, Betancourt reached the conclusion that its popular distribution was already exceeding the limits of a reasonable increase in consumption levels:

> [...] it was a matter of concern for the Government team to see how, under the aegis of the petroleum bonanza, the people merrily spent [...] the money which came into their hands [...].
>
> And in this matter of not saving at all [...] the Venezuelan was typically representative of mining communities. The country as a whole was an imitation of the California of the days [...] of "madness of gold". To spend until pockets were empty was a sort of national slogan [...].
>
> The most favored sectors by the lottery of petroleum dilapidate the money that has so easily come by [...] with the merry irresponsibility of any mining clique. The contagion has spread to the rest of the population, and especially to the middle classes (Betancourt, 1956, p. 456f).

2.3.3. Uslar Pietri

According to Arturo Uslar Pietri writing in 1948:

> Venezuela before petroleum was [...] a country that lived off its work [...].

> With petroleum [...] the increasing wealth it is going to have does not come from Venezuelan labor (Uslar Pietri, 1948, p. 213f).

Now the country was living off the consumption of 'a capital deposited by nature in the subsoil', instead of reinvesting it. The policy of distributing petroleum fiscal income to the population at large, between 1945 and 1948, he considered to be disastrous:

> The true harm, the almost irreparable harm, [...] lies in that [...] the sense of economy in the Venezuelan populace [...] has become perverted. At all social levels, it has been taught to disdain toil for manna, to think in terms of magic rather than accounting, to have no notion of prices, costs and economic equilibrium [...].
>
> The notion has been lost of what a normal economic life is, and it has become almost impossible to return to it (Uslar Pietri, 1948, p. 117).

Uslar Pietri was also opposed to the dominant role of the state in the industrialization process. In 1958, referring to the whole of the period between 1945 and 1957, he left no doubt that such role should correspond to the private sector:

> A large part of that wealth, in the past, has been invested in creating a State Capitalism which, surely, is the most extensive and powerful in the world this side of the Iron Curtain. This State Capitalism has grave consequences. If it continues to grow indefinitely, Venezuela will become a country no longer made up of dependents on petroleum but of dependents on the state, and this monstrous State Capitalism will fatally become, as it has become in the past, a terrible machine for tyranny. We should make a real effort so that in the immediate future, a large part of that wealth that will go through the hands of the state, does not end up by swelling, strengthening and extending that State Capitalism [...] but instead is devoted, purely and simply, to strengthening the middle class and to really creating a capitalism within Venezuela, which in our days is barely giving off its first birth cries and which, perforce, will be a stabilizing, liberal and democratizing force in the immediate future of the country (Uslar Pietri, 1958, p. 13f).

He did not doubt the suitability of public ownership of petroleum, which he called an "almost providential" fact that "came to serve the interests of Venezuela" (Uslar Pietri, 1948, p. 39). However, as far as petroleum rent and its distribution were concerned, he took refuge in a historic fiction:

> It would be possible to imagine a different process [...] whereby the owners of the surface would have owned the subsoil too, in which the petroleum wealth would have been distributed regionally, and would have ended up in the hands of private parties to start with, and in which those private parties would have been the entrepreneurs and creators of Venezuelan capitalism and the state would have partaken of this wealth solely through taxation, receiving its share

> from the individuals and corporations which would have been its owners. But that is not how things went (Uslar Pietri, 1948, p. 51f).

Thus, in its origin, public ownership of petroleum was most advisable, but in its destination, it was most suitable that the investment of the IPR be in the hands of the private sector. Definitely, it was about its *privatization*, although for the sake of appearances, the flows of money would be characterized as loans, which, as a matter of facts, would never be paid. Thus, he did not have to explain nor to justify that actually it was about privatizing public income. But he left no doubt that the population at large should only obtain benefits from petroleum rent in their capacity as wage earners, once it had been transformed into productive investment.

3. Control of Production and Conservation

Betancourt brought to a close his exposition (in 1956) about the 'sowing the oil', between 1945 and 1948, with the following reflections:

> In 1948, the economic and financial situation of the country presented prospects which in appearance were very favorable. The increase in production was being achieved; the budgeted expenditures of the state were substantial [...]; and the accumulated international reserves in the [BCV] were very substantial and gave strong support to the currency (Betancourt, 1956, p. 383).

However, this was mere 'appearance', given that:

> Behind this very favorable panorama there was [...] a reality which was impossible to ignore: the excessive dependence of the Treasury on the rising post-war petroleum production, true fuel and lubricant of the Venezuelan economy (Betancourt, 1956, p. 452).

Betancourt justified the dizzying rise in production during his government as a sacrifice undertaken by Venezuela to collaborate with the reconstruction of Europe, even though the rapid depletion of the mineral wealth in the national subsoil was advisable neither for Venezuela nor for the American continent at large:

> We made this known to the companies, warning them that once the heightened demand of the post-war period had been covered, reasonable limits to production would be set (Betancourt, 1956, p. 242f).

Betancourt suggested that the country might, within the context of the Hemispheric Treaty, become an integral part of the system for the control of production managed by the United States, on the one hand, and would curb production in order to reduce the 'excessive dependency' of the Venezuelan public purse on the international petroleum rent, on the other hand.

Furthermore, such control would be legitimized by being presented under the guise of a *conservationist* policy:

> The idea that nations should zealously take care of their natural resources is relatively novel. There has arisen a contemporary school which already has its apostles and prophets, animated by a conservationist mystique. Mankind begins to realize that its forests, lands, waters cannot be used without limits. Reserves have to be conserved for the future and natural riches have to be exploited in a rational fashion. This new conception of the exploitation of the natural endowment of a people becomes heightened in the case of non-renewable resources. Among the most characteristic of these is petroleum. It is a wealth that becomes exhausted (Betancourt, 1956, p. 271).

4. National Accounts

From the First World War onwards, the public budgets of the developed capitalist countries acquired increasing importance. After the world economic crisis of 1929, there developed in the Anglo-Saxon sphere a system of National Accounts, which was to sum up the macro-economic data of relevance for the design of the national economic policies of the state. Of particular interest to us is the Gross Domestic Product, especially as it relates to national income. *GDP at factor costs* consists of *Wages and Salaries* and the *Operating Surplus* – both expressed in gross terms – *before* any direct taxes due (which, as a practical matter, are income taxes), on the one hand, and the *Consumption of Fixed Capital*, on the other hand. These three components add up to the *GDP at factor costs*.

As can be seen, the GDP at factor costs makes no reference to ground rent, as Economic Science considers that ground rent is always a differential rent, subordinated to capital and irrelevant to prices. Not surprisingly then, National Accounts provides no place either for the Ricardian conception of minerals underground as deposits of 'valuable commodities'. In the words of Richard Stone, one of the creators of the National Accounts System:

> We do not pay nature for accumulating large deposits of coal in the earth [...]. Coal underground is worth nothing *per se* [...] (R. y G. Stone, 1961, p. 14ff).

However, the state also intervenes in the economy through *Indirect Taxes and Subsidies*, which have the deliberate aim of modifying prices. This is the case, for instance, of excise taxes on alcoholic beverages, whose consumption is meant to be discouraged, or with subsidies to milk, whose consumption is meant to be encouraged. Adding Indirect Taxes and Subsidies to GDP at factor costs results in *GDP at market prices*, the total GDP.

However, there is something else. As we have seen previously, the public ownership of oilfields constitutes nationally a free gift of nature, but internationally it constitutes a source of international rent. However, the

National Accounts do not differentiate between production for the domestic market, and production for exports.

But the United Nations, created in 1945, was to promote this system of National Accounts in all its member countries. Consequently, in countries which export raw materials, any EIS is simply included as part of the GDP of the corresponding sector, like all operating surplus.

4.1. Venezuela: Gross Domestic Non-Rent Product

National Accounts were introduced to Venezuela in 1950, in compliance with the system promoted by the United Nations, and were entrusted to the BCV. Table V.1 shows the GDP figures for 1968, albeit condensed to only two sectors: petroleum and non-petroleum. Hence, Petroleum GDP (GDPPet) shows an operating surplus of 8.687 million Bs.

Table V.1

CENTRAL BANK OF VENEZUELA
GROSS DOMESTIC PRODUCT 1968

Million Bs	GDPPet	GDPNPet	GDP
Wages and Salaries	1,201	18,898	20,099
Operating Surplus	8,687	9,328	18,015
Consumption of Fixed Capital	963	3,881	4,844
GDP at factor costs	*10,851*	*32,107*	*42,958*
Indirect Taxes - Subsidies net	109	2,406	2,515
GDP at market prices	*10,960*	*34,513*	*45,473*

However, the Central Bank, by including the EIS, overestimates substantially the GDPPet, *as it is not part of the national GDP properly understood, but reflects part of the GDP of the importing countries*. Baptista and Mommer corrected the United Nations scheme (Baptista and Mommer, 1986), referring to the result as non-rent GDP (GDP(NR)), as it excludes the EIS from the petroleum GDP estimated by the BCV. In 1968, the GDPPet(NR) was only a third of the GDPPet estimated by the BCV, which means that GDP(NR) ends up being substantially less than the GDP(BCV). However, the EIS of the petroleum sector was estimated to be 7,332 million Bs. (Table V.2)

Table V.2			
BAPTISTA AND MOMMER **NON-RENT GROSS DOMESTIC PRODUCT 1968**			
Million Bs	**GDPPet(NR)**	**GDPNPet**	**GDP(NR)**
Wages and Salaries	**1,201**	**18,898**	**20,099**
Operating Surplus	**1,355**	**9,328**	**10,683**
Consumption of Fixed Capital	**963**	**3,881**	**4,844**
GDP(NR) at factor costs	***3,519***	***32,107***	***35,626***
Indirect Taxes - Subsidies net	**109**	**2,406**	**2,515**
GDP(NR) at market prices	***3,628***	***34,513***	***38,141***

Extraordinary International Surplus:	***7,332***

4.1.1. GDP(BCV) vs. GDP(NR)

Let us consider the most important consequences which derive from the use of a system of National Accounts which not only subordinates land to capital from the national point of view but also from an international standpoint.

Income Distribution by Factors of Production

In 1968, according to the BCV, income distribution by factors of production – the ratio between Wages and Salaries, on the one hand, and the Operating Surplus, on the other – was 53:47 (Table V.1). By international standards, this would have been an exceptionally unfavorable share for labor. However, this ratio is explained because the whole of the EIS is included in the operating surplus of the petroleum sector. But if we look only at the non-oil sector, the income distribution by factors of production was at a 67:33, which is perfectly normal for a country at Venezuela's development level. Thus, the seemingly 'disadvantageous' income distribution for labor is rooted in the petroleum sector, where the ratio was 12:88 in favor of capital. Nevertheless, in both the domestic and international literature on the Venezuelan economy, one customarily finds the assertion that this is a country with an exceptionally unfavorable income distribution for labor (B. Mommer, 1986).

However, looking at GDP(NR), the income distribution ratio is 65:35 in favor of labor, while in the petroleum sector the income distribution is 47:53 in favor of capital (Table V.2), which reflects the exceptional capital intensity of the sector.

Multipliers of Petroleum as a Productive Activity

The degree of integration of a specific productive activity within the national economy is usually measured by means of an input-output matrix, and referred to as the Leontief multiplier. If this procedure is applied to the petroleum sector, on the basis of the National Accounts of the BCV, the result is that the petroleum sector displays a very low multiplier effect, which means that producing one additional dollar's worth of petroleum has a very modest induced demand effect on other sectors of the national economy. This has given rise to the widespread and mistaken notion that the hydrocarbons sector represents an 'enclave' within the national economy. The problem lies in the fact that the price of hydrocarbons includes a rent element which, evidently, has *per se* a multiplier of zero, since the natural resource itself does not have a cost of production.

However, if the GDP(NR) is considered including the productive activities as such, with a national operating surplus of 10% on the net capital stock but excluding the extraordinary international surplus, the Leontief multiplier calculated on this basis reveals a higher degree of integration of the petroleum sector with the national economy, compared to the average of other industrial activities. This is hardly surprising, as the non-petroleum sector developed, throughout the decades, around the petroleum sector. Even if machinery and technologically sophisticated equipment were always imported, the local capabilities to cover extensive infrastructure works, as well as all kinds of services and inputs, gradually developed within the national territory (B. Mommer, 1993).

Remittances of Foreign Capital

BCV National Accounts do not quantify the EIS, whereas they do quantify the remittances of foreign capitals, in both the petroleum and the non-petroleum sectors. These remittances, in 1968, came to 3,560 million Bs. Their magnitude, compared with the national economy, placed Venezuela as the Latin American country most intensely exploited by foreign capital. But that same year the Venezuelan state had benefited from an IPR of 6,432 million Bs, and the remaining 900 million Bs (= 7,332 – 6,432) had accrued to the benefit of the international oil companies as extraordinary profits. But taking into account the IPR as a unilateral international transfer, the balance in favor of the country was positive, highly positive. From this perspective, Venezuela actually appeared as the most privileged country in the whole of Latin America.

Historical Series

In order to be able to evaluate and compare the performance of a national economy over the years, the National Accounts have to use *base years*, which are years when a more exhaustive and detailed investigation is undertaken.

Thereafter, the yearly data at current prices, and also at constant prices, are based on samples. In this way the impact of inflation and merely circumstantial variations of prices are neutralized, allowing the measurement of the growth rates of the national economy and its different sectors. However, economies also vary over the years – quantitatively and qualitatively –, due to technological development and patterns of consumption. This is why the base year is changed every ten to fifteen years. Finally, the series of past periods are spliced with the new base year to obtain long-run series expressed in constant prices.

The year 1968 was a base year for the national accounts of Venezuela, and the next base year was 1984. However, it makes no sense to present statistical series of the petroleum sector at constant prices, given that the export prices in question have been, and continue to be, *variables*, moreover of major political and economic importance, not only nationally but also internationally. The GDPPet(BCV), at constant prices, simply varies in accordance with the volumes produced over the period to which any specific base year applies. Yet in 1968, the basket of exported hydrocarbons had a price of 6.29 US$(1997)/barrel, but in 1984 the price was 32.32 US$(1997)/barrel, a difference that made impossible any reasonable splicing on the basis of constant prices.

In contrast, the scheme proposed by Baptista and Mommer, GDPPet(NR) expressed in constant terms, does make sense. And as far as extraordinary international surpluses are concerned, the necessary adjustment has to be made by means of the *price index, in dollars, of goods imported by Venezuela*; that is to say, the series has to be deflated in accordance with the country's international purchasing power. In sum, the national petroleum sector cannot be sensibly described on the basis of one variable only; as a minimum, two variables are necessary: GDP(NR) and EIS (Graph V.8).

Figure V.8

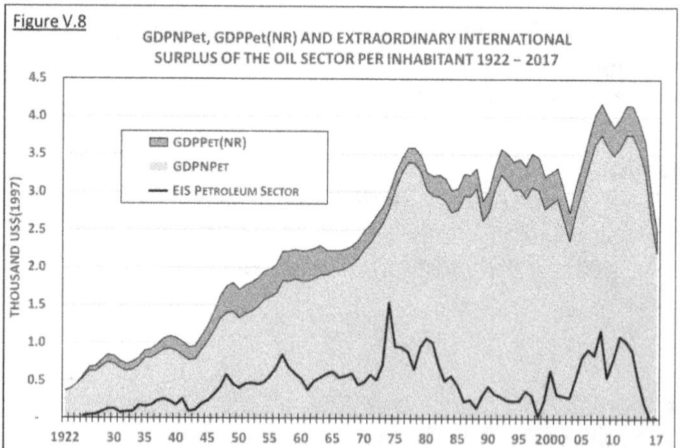

4.2. Conclusions

With the development of capitalism in Venezuela, the notion that oilfields were natural capital became entrenched, and international petroleum rent was considered to be a measurement equivalent to depreciation. Hence, its ultimate destination could be none other than its re-investment fully in development of the non-petroleum national economy. In 1966, Uslar Pietri, for instance, estimated at 32 billion US$ "the capital that Providence has given us so that we, intelligently, may build a country, and it is our duty to do so" (Uslar Pietri, 1966, p. 73f).

In contrast, attributing the origin of the IPR to the exploitation of labor in the petroleum industry posed the problem that, aside from the fact that its workforce was the best remunerated in the country by far, the sheer magnitude of the international petroleum rent led to absurd results (Mejía Alarcón, 1972). The policy of popular distribution of the IPR would finally come to be justified in the name of the formation of human capital.

Once Betancourt and Pérez Alonzo had to face the fact that the IPR was already exceeding the absorptive productive capacity of the country, they turned to conservation as the raison d'être of the landlord state. Thus, the idea would be the defense of the natural resource for the benefit of future generations.

But the conception of Peltzer of IPR as a unilateral transfer that bore no relation to any domestic productive effort was simply unacceptable and even unimaginable to Venezuela's political elite. Mayobre himself, who would be Finance Minister in a number of governments, never again made any reference to his thesis from 1944.

Moreover, in the National Accounts, the BVC presents the IPR as part of the operating surplus of the petroleum sector, thereby foreordaining its use for investment purposes. Hence, the BCV's National Accounts do not fulfil their most elementary purpose, which is to provide the government with the information necessary to design a sensible economic policy, as they systematically ignore the fact that Venezuela is an oil country

At the same time, the whole debate surrounding the issue of 'petroleum and development' obviously made reference to petroleum as a source of international ground rent, and not as a productive activity. This is why this debate, as a practical matter, ended up being entirely about 'petroleum income and non-petroleum development'. The lack of sensible historical time series on GDPPet(BCV) in real terms was 'solved' at a national level by limiting all debate and discussion to the non-petroleum sector, identified with the 'true' Venezuela. The GDPPet(BCV) series, at constant prices, made sense to no one, not even to economists.

For its part, the academic world continues to ignore the concept of GDP(NR). But imposing an accounting scheme created to serve non-petroleum countries means that, for a country such as Venezuela, it becomes a straight-jacket. As ever, when reality and economic theory clash, it is always reality that has to change. The conclusion drawn from these national accounts was that the Venezuelan economy was deformed; that in order to fit better within the confines of its straight-jacket, it would somehow have to stop being an oil country. Moreover, from the Second World War onwards, generations of Venezuelan economists and social scientists were formed in the country's universities, on the basis of a syllabus conceived for non-petroleum countries.

5. International Petroleum Rent and Economic Growth

Given the high petroleum fiscal income, the most backward national productive activities were abandoned and only the more productive sectors were maintained, thus liberating the corresponding workforce, and also attracting a qualified immigration, which were both necessary for building the most varied kinds of infrastructure. Thus, an extraordinarily rapid urbanization process took place, without there being a previous development of the agriculture and stock breeding sectors. At the same time, wages and profits rose, backed by imports. The national economy grew thanks to the mere fact of its restructuring, and this restructuring was paid for by the government using the IPR, which meant that it entailed no costs to the beneficiaries of the new infrastructure.

5.1. Import Substitution

The growth of the national economy, beyond its reorganization on the bases of the IPR, would require the development of national, non-petroleum, productive forces. In practice, this economic policy would come to be known as 'import substitution'.

However, it is necessary to distinguish between the investment originating from the national surplus, on the one hand, and the investment originating from the IPR, on the other. The former not only creates a quantum of supply but also a demand of the same magnitude, in accordance with what is known as Say's law. This supply/demand quantum is equivalent to the sum of capital depreciation, wages and salaries and profits. In contrast, the national investment originating from the IPR and aimed at substituting imports does not comply with Say's Law, as it does not give rise to a corresponding demand, since the natural resource as such has no production cost. As a consequence, the amortization of the investment in question requires not only that semi-finished goods continue to be imported, but it also becomes necessary to distribute among national consumers, over the years, an equivalent amount to the depreciation of capital. This is the only way in which a corresponding demand can come into existence, and it is also the only way for investors to recoup their capital and to invest it again. Hence, even if over the years some of the imported inputs might be produced locally – as a result of the development of productivity –, it would definitely still be necessary for there to be a balanced distribution of the IPR between the portions destined for investment and consumption, respectively. The investment of funds coming from the IPR has no effect whatsoever on the fact that the whole of the nationally distributed IPR would continue being turned into imports.

5.2. Import Substitution and Exports

The inward investment of the IPR was faced with a small domestic market. Inevitably, the development of productivity would require mass production and, in consequence, once the national demand was met, a significant part of output would have to be exported. Thus, Betancourt, for example, made reference to the possibility that 'manufactured goods which might be produced in the country' would 'find an outlet in the vast Latin American market'.

However, the distribution of the IPR made production more expensive, and would therefore obstruct the competitiveness of non-petroleum exports. In the new industries that sprang up as a consequence of the economic policy of import substitution, the distribution of the IPR was reflected in higher wages. In addition, domestic investors, with high import tariffs as a foundation, also aspired to realize higher profits, beyond levels which could be justified by differences in productivity. That is, even if the productivity of these industries were to reach international levels, their output would still be costlier to produce, because production costs would include a component of distribution of the IPR.

Ultimately, when the national entrepreneurs, in the process of capital accumulation, had to confront the limits of the national market, it made no sense for them to continue to invest with a view towards export markets. Consequently, the national entrepreneurs ended up converting the surplus in question into foreign currencies.

5.3. Conclusions

With the development of capitalism in Venezuela and the continuous growth of the IPR, the idea of *sowing the oil*, as it had been spontaneously unfolding, ran into serious limitations. Of course, it was perfectly possible to distribute the IPR among private citizens, without thereby implicating the production process in any way. In fact, there always existed mechanisms to do just that; for example, in health and education. Also, high wage levels could be substituted by monthly bonuses for all Venezuelans residing in the country. In this way, profit levels could be limited to 'normal' levels, although they would be subject to a minimal tax burden.

However, at no point was this problem as such put up for discussion in order to find a solution aimed at making sure that the mechanisms for distributing the IPR would not adversely impact production costs. Instead, such political discussion as there was focused on a polarized clash pitting on the one hand the argument (put forth by the business community and its spokespersons) that the lack of competitiveness stemmed from high wages in real terms, on the other hand, the argument (put forth by organized labor and its spokespersons) that its cause lay in the monopolistic practices of entrepreneurs. For their part, Betancourt and Pérez Alfonzo, as representatives of the landlord state ended up seeking refuge in the discourse preaching the need for conservation of an exhaustible natural resource.

In any case, the success of the *sowing the oil*, understood as an accelerated growth of the national economy, would inevitably lead to a national economy with growth rates that – in the best of cases – would be comparable to those of other countries around the world (in accordance with their respective levels of productivity) albeit still characterized by an exceptional quality of life.

6. Petroleum and Dependence

6.1. Fiscal Regime

According to Pérez Alfonzo:

> The petroleum policy of democratic governments had always taken into account the *business conditions* which characterize the relationship between Venezuela and the international petroleum industry. [...] this relationship [...] modifies the character of the concessionaires as mere tax payers, and adds to it their role of administrators, on behalf of the state, of the mineral resources under exploitation. [...] The [idea of] *reasonable participation* meant that Venezuela should benefit from the exploitation of the petroleum concessions, taking the remainder of all profits realized after the deduction in favor of the concessionaires of those quantities which, also reasonably, were due to them as an adequate remuneration for the capital invested [...] (Pérez Alfonzo, 1971, p. 56; emphasize in original).

Therefore:

> It is mandatory to recoup from the concessionaires the highest possible income, but respect at the same time the right of investors to get a reasonable remuneration for the capital invested in the exploitation of the resource. Unlike the relationship with regular tax payers, where fiscal needs are the justification for the payments by private parties, in the oil business the concessionaires have to turn over to the fiscal authorities the whole of the surplus, irrespective of the greater or lesser urgency of public expenditure (Pérez Alfonzo, 1971, p. 58).

For Pérez Alfonzo the existence of a proprietorial rent was inconceivable. The aim of the state was to be collecting excess profits from an industry which he considered to be a public service. However, in 1966, the Bill reforming the Income Tax law, concentrating on taxing excess profits, faced another resounding *no* of the great international oil companies, in this case Creole (Exxon), Shell de Venezuela and Mene Grande (Gulf). For Pérez Alfonzo, this was one 'no' too many:

> The exercise of sovereignty […], would run into ever greater difficulties if these economic forces were to continue uncontrollably taking root on the fringes of the political-legal system of the country.
>
> […] What has happened has to be evaluated in all of its ramifications in order to allow making decisions aiming at moderating and correcting the excessive influence which all Venezuelans could realize. Of course, what is most urgent is to plan for […] the gradual albeit inflexibly consecutive diminution of foreign investments in petroleum, as well as of the US-investments in general. The characteristics of investments from that country in Venezuela, the magnitudes that they have reached, the habits of domination and exploitation which are its hallmark […], all come together to make dangerously undesirable everything which might keep such investments from declining from the exorbitant position which they have achieved in Venezuela (Pérez Alfonzo, 1967, p. 107f).

However, Pérez Alfonzo said not a peep, even in hindsight, about what the 1943 Petroleum Reform had represented. Pérez Alfonzo would continue to disqualify it, making reference to the fact that the international oil companies had been compelled to accept Venezuela's tax sovereignty as one among the "many *apparent* advantages […] which every legal system recognizes without there being any need for special acceptance, as in the case of the ballyhooed submission to the taxing powers of the state" (Pérez Alfonzo, 1971, p. 11f; emphasis ours).

6.2. Capital Absorption Capacity

Moreover, Pérez Alfonzo now made reference to the 'limits to making good use' of petroleum fiscal income, which had already been surpassed during the period 1945 to 1948. He also stressed that, in the non-oil sector and non-residential construction, investment had lagged behind in relation to petroleum fiscal income from as far back as 1950, and that the country had in this way become increasingly 'dependent' on petroleum. Indeed, with the increase in both export volumes and prices, petroleum fiscal income grew to the extent that, from 1960, it exceeded investment in non-oil sector activities and non-residential construction (Figure V.9).

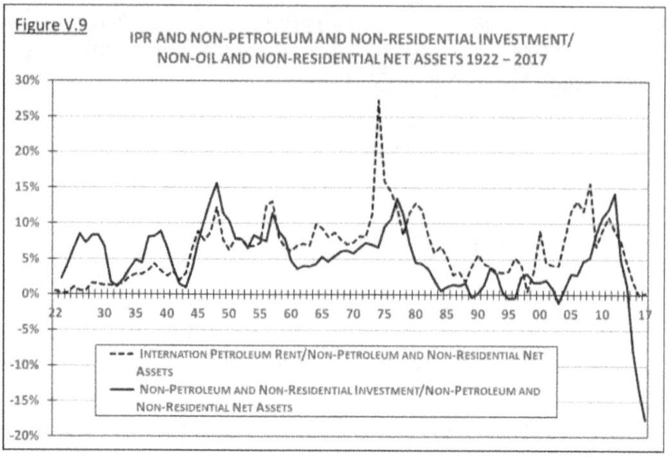

Now Pérez Alfonzo presented the success of the national petroleum policy as regards petroleum as a source of an international ground rent in the negative light of a rising dependence:

> The principle, otherwise obvious, was demonstrated that there exists a relatively fixed capacity to employ new capitals usefully. As soon as that capital absorption capacity is exceeded, economic growth ceases to correspond with greater amounts employed beyond such limits (Pérez Alfonzo, 1971, p. 18).

Indeed, the capital absorption capacity of a country, generally speaking, is proportional to its productivity: to produce is to invest, and to invest is to produce (Baptista and B. Mommer, 1985).

Hence, Pérez Alfonzo suggested that production be reduced because "so long as we allow this natural resource which is irremediably running out, to be produced at a rate leading to its disappearance, which is so fast that the speed at which income increases is greater that the increase in growth in the rest of the economy", the result would be "a greater dependence towards petroleum, at the same time as its extinction was being hastened" (Pérez Alfonzo, 1971, p. 18f).

Prudently, Pérez Alfonzo only proposed reducing this 'dependence', and the non-oil economy would have to grow at a per capita rate considered normal, of between one and two percent per inhabitant (Pérez Alfonzo, 1971, p. 41), but the IPR could still grow, albeit at a lower rate. In consequence, even if part of this IPR were consumed, dependence itself would be shrinking.

According to Pérez Alfonzo, from 1959 onwards and amid difficult conditions, the Betancourt government had initiated an austerity policy by reducing public expenditure and paying of the debts left by Pérez Jiménez. But after the devaluation of the petroleum bolívar in January 1964, the subsequent Leoni (AD) and Caldera (Copei) governments had gone back to increasing it disproportionately. In support of his thesis, Pérez Alfonzo pointed to "the incredible capital accumulation which some Venezuelans keep abroad", a fact which constituted "tangible evidence of the surplus capital that swamps the capacity of employing it with efficiency in Venezuela" (Pérez Alfonzo, 1971, p. 109). Even more, Pérez Alfonzo ended up by agreeing with the "renowned intellectual and politician Arturo Uslar Pietri" who, as far back as 1946, from his exile in New York, had warned about the danger of a Venezuela "dependent" on oil, and had denounced the "*lightness of the fiscal burden* in Venezuela" (Pérez Alfonzo, 1971, pp. 19 and 48; emphasize in original):

> [...] the bureaucracy meets the rising need for collective services by liquidating national assets, which allows it to avoid facing the always difficult task of collecting from those who ought to contribute on a larger scale.
>
> [...] the big taxpayers applaud and stimulate Public Spending which is not funded by their own contributions and which, consequently, works to increase their own Private Assets (Pérez Alfonzo, 1971, p. 50).

In particular, what was happening was a:

> [...] direct transfer of wealth from the *common patrimony* of all Venezuelans to the *private patrimony* of a few privileged ones, always on the hunt for unjustified benefits (Pérez Alfonzo, 1971, p. 54ff; emphasize in original).

6.3. The new Juncture of Ground Rent

In 1960, the country's proven reserves had reached their apex, at 17.4 billion barrels, but by 1970 they had decreased to 14 billion barrels. These figures corresponded to the areas granted in concession and *explored* by the concessionaires, and it was not at all clear what the decrease in proven reserves could mean for the country. What is certain is that, within OPEC, the Venezuelan government continued to put forward a scenario predicated on a continuous growth of production.

In 1970, nobody in Venezuela was aware that the country was reaching its maximum historical output. Moreover, the country harbored no doubts that any drop in output would be followed in lockstep by petroleum fiscal income. This was the only conceivable scenario and corresponded to the eternal threats made by the concessionaires. The difference was that Pérez Alfonzo considered the drop in output as a positive development. When this happened, followed by an increase in petroleum fiscal income, he insisted on a further reduction in exports. Finally, with the explosive increase in the IPR towards the end of 1973, he proposed that, in 1974, output be reduced to 1.5 million barrels/day. That is to say, in the same measure as the income per exported barrel grew, output would have to be reduced (Pérez Alfonzo, 1976, p. 170).

Moreover, the bonanza of petroleum fiscal income should be dedicated to liquidating the remaining public debt, and also to pay the compensation corresponding to the nationalization of the iron industry, of the petroleum industry and, in general, of all foreign capital of any importance (banks, insurance, supermarket chains and others). These payments of compensation would have to be liquidated in cash, something which was perfectly possible given the extraordinary magnitude of petroleum fiscal income.

6.4. Conclusions

Once he had retired from his official political life, Pérez Alfonzo presented himself before the national public opinion as the embodiment of the landlord state. The crucial point of his petroleum policy was that the IPR should lose importance relative to non-oil GDP, so that 'dependence' would be thereby lessened.

Moreover, Pérez Alfonzo insisted that petroleum fiscal income:

> For the most part originated in the intrinsic value of the liquidated non-renewable product, and not from the economic activity deployed to take it to the markets. As has been very well stressed, these are monies which are not won through work, or entrepreneurship and, therefore, they are monies which have not served to prepare for giving them a destination that is appropriate and useful for society.
>
> There is something that makes the situation even worse. Aside from the fact that oil money neither prepares for, nor teaches, how to invest it judiciously, it is foreign money. The exported oil becomes foreign currency which can only be used for the acquisition of goods and services outside the national territory (Pérez Alfonzo, 1976, p. 178f).

As the personified landlord state, Pérez Alfonzo now indeed attributed an 'intrinsic value' to oil, and IPR could only legitimize itself in terms of its destiny, in its productive investment. Beyond the country's capital absorption capacity, the international petroleum rent could not be legitimized. For the time being, its consumption had to be tolerated, but on the proviso that it would shrink over time. Pérez Alfonzo's conclusion was to simply say 'no' to Venezuela as an oil country:

> *Sole Alternative: Drastic Reduction in Output* (Pérez Alfonzo, 1976, p. 232; emphasize in original).

He suggested that an end be put to Venezuela as an oil country – so to speak – through the back door, that is to say, that Venezuela would no longer be a petroleum exporting country of importance. According to Pérez Alfonzo, Venezuela as an oil country had no future; the 'true' Venezuela was the non-oil Venezuela.

7. The New Model of Development

For the Venezuelan political world, the legislative process between 1970 and 1973 revolved around the recovery of full tax sovereignty and reversion. Tax sovereignty was indeed recovered in 1970, when the National Congress authorized the National Executive to fix, unilaterally and sovereignly, the fiscal reference prices. Regarding reversion after the expiration of concessions, the point of view of the National Executive about this issue prevailed with the Reversion Law, enacted in August 1971.

This legislative process took place in the context of a rising petroleum fiscal income, despite the substantial decrease in output, something which by itself strengthened the Venezuelan position in connection with the reversion that was approaching. Then, towards the end of 1973, the explosive increase in international petroleum rent came about and, with it, a further extraordinary strengthening of the Venezuelan position. Carlos Andrés Pérez, by then already President elect – the first of a new generation of political leaders – proceeded to implement what he thought would be a new model of development.

7.1. Gumersindo Rodríguez

Shortly after he had assumed the Presidency, Carlos Andrés Pérez referred to the origin of international petroleum rent identifying it, as was usual, as a circumstantial wealth that:

> [...] arises from the impoverishment of the nation due to the exhaustion of our extractive wealth. It is like a loan which the Government takes against the wealth which corresponds to future generations. But, in truth, [...] we are incurring in an expropriation (Quoted in Gumersindo Rodríguez, 1979, p. 86).

Hence, the new government would seek to put an end to this plundering of future generations, making sure that this 'circumstantial wealth' was turned into productive investments. The key ideologue of the new development model was the Planning Minister, Gumersindo Rodríguez. Regarding the models of the past, he referred to "Recent Errors in the Planning for Development", not only in Venezuela but throughout Latin America:

> The economists of previous generations who have preceded the younger graduates have constituted one of the most fragile sources for the defense of the intellectual sovereignty of our peoples. Through their influence, there have been absorbed models of the conduction, control, regulation or planning of our economies which do not correspond to our spatial, social and economic realities; and which, upon being implemented, act as additional structural factors of distortion (Gumersindo Rodríguez, 1979, p. 92f).

In particular, he referred to the policy of 'import substitution'. This new generation of 'sovereign intellectuals', to which Gumersindo Rodríguez in his own judgement belonged – with his graduate and postgraduate studies in Economics in Great Britain – had decided to impose a new model of development:

> Our economies are radically different: They are fractured in their productive structure and are composed of a modern corporate sector, through which multinational enterprise acts as the technological and management leader; and a marginalized or subordinate economy in the countryside and the cities, which is nothing more than an automatic device in our societies for the vegetative retention of legions of apparently occupied workers. These two economies lead parallel lives, and when they come into contact, it is only so that the better organized and planned economic units restrict and impoverish the marginal economies through the appropriation of their scant economic surpluses (Gumersindo Rodríguez, 1979, p. 95).

He alluded thereby to the economic theory put forward by Raúl Prebisch, with regard to the deterioration of the international terms of trade of agricultural products in Argentina, in particular, and in the countries exporting agricultural products, in general. Prebisch considered that the deterioration in these terms of trade was due to the decline of salaries of these countries (D. Mommer, 1977, 1st Part). For his part, Gumersindo Rodríguez pretended to attribute the extraordinary profits of the concessionaires in Venezuela to an inequitable distribution of the national income. As a matter of fact, the government immediately decreed substantial wage increases and, on the basis of an output of three million barrels per day, public spending doubled. But there was much more than this at stake, because "New Paths for Venezuelan Planning" were to be defined:

> More than planning, what is in play is a strategy to modify the structure of society and open the way to a Venezuelan economic system founded [...] upon a more just distribution of [...] income [...], in an industrialization that puts emphasis on the production of machinery and on the rational generation and adaptation of technologies, and a rational exploitation of our natural and human riches (Gumersindo Rodríguez, 1979, p. 96).

It would no longer be a matter of substituting the importation of consumption goods for semi-finished ones; instead, the very root of the problem would be attacked, importing capital goods. Supposedly, it would be a matter of rectifying the wrong turn taken in the past (Gumersindo Rodríguez, 1979, p. 96ff). According to President Carlos Andrés Pérez:

> A National Plan is being drawn up with a strategic perspective which will transcend the temporal horizon to encompass several constitutional periods, delimiting the boundaries of the economic action of the state and private enterprise, apart from the changes to achieve a fair distribution of income (Quoted by Gumersindo Rodríguez, 1979, p. 136).

In order to guarantee the continuity of investment projects and create what Carlos Andrés Pérez would christen as the 'Great Venezuela', the Organic Law of the National Public Exchequer was modified in order to allow for an exception to the doctrinal principle of the Unity of the Treasury. In this way, about half of the 1974 fiscal petroleum income was to be deposited in an Investment Fund, under the direct control of the Executive, precluding any involvement on the part of the Legislative Power.

Towards the end of 1974, the government nationalized the iron ore industry together with its concessions, in addition to nationalizing other important foreign investments – or, at least, acquiring majority shareholdings. At the same time, it announced all kinds of megaprojects. An automotive as well as a tractor manufacturing industry was going to be created; the steel and iron making industry would be expanded; the petrochemical industry would be further developed and a national railroad network would be built. Evidently, in view of their magnitude, all of these would be state projects, although state enterprises were also to be reformed accordingly in order to subject them to the profitability criteria of the private sector, including the right to become indebted. The traditional figure of the autonomous institute disappeared, thereafter to be limited to entities entrusted with strictly administrative tasks, and no longer to business activities.

This new development model – which, in 1976, was enshrined in the V National Plan – was to go hand in hand with a process of indebtedness. The compensation payments corresponding to the nationalizations were to be paid by instalments for a period of 10 years. But already in 1976, the National Congress authorized the government to contract debt in an amount of 60 billion Bs (14 billion US$). This policy of massive investment would not only tie down the current high IPR, but also the future IPR (B. Mommer, 1989).

Moreover, the policy of investing in the midst of an extraordinary financial bonanza also went hand in hand with the significant loss of foreign and national managers who were not willing to work for state enterprises. Thus, the financial boom notwithstanding, there was a loss in human capital and, therefore, a decrease in the capital absorption capacity of the national economy, which was already swamped anyway. The inevitable failure of this policy began to be felt starting in 1975, with the contraction in the gross profit rate of private non-petroleum capital. But the remuneration of labor would continue to increase until 1978, which was Carlos Andrés Pérez's last year in office (Figure V.10).

Figure V.10 YEARLY WAGES AND SALARIES/EMPLOYEE AND NON-OIL OPERATING SURPLUS/NON-OIL NET ASSETS 1950 - 2017

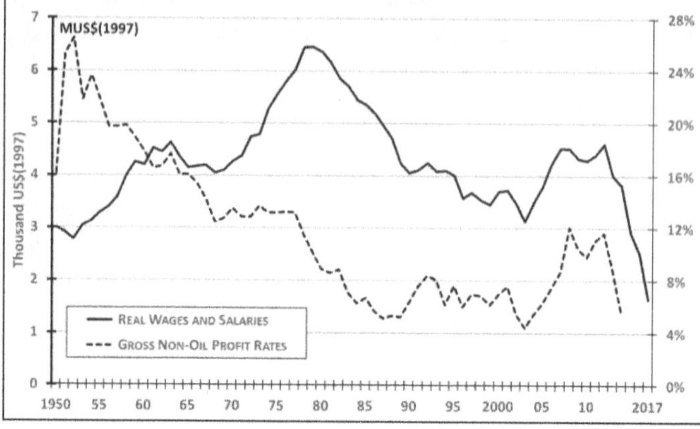

As a matter of fact, between 1975 and 1979, the levels of investment in the non-residential construction of the non-petroleum sector surpassed all records of the past. Furthermore, this was only made possible by a boom of petroleum as a source of international ground rent; but petroleum as a productive activity contracted and, more generally, so did human capital, with the nationalizations of foreign investments. Furthermore, with the exit of foreign concessionaires, national businesses lost an ally of great importance not only on the political but also on the economic side and in relation to the development of their own entrepreneurial capabilities.

7.2. Foreign Public Debt

Pérez Alfonzo harbored no doubt that:

> For the people who have squandered what is theirs, giving off clear signals of waste, indebtedness means ruination, destruction. Hence, the trend evidenced by the V National Plan allows one to characterize it as the National Destruction Plan (Pérez Alfonzo, 1976, p. 369).

However, in 1978, having already left the Planning Ministry, Gumersindo Rodríguez made reference to the government's decision to "contract international loans denominated in dollars [...] in order to ensure the financing of the investment programs of the public sector", and pointed to Pérez Alfonzo as "the staunchest enemy of this program of foreign indebtedness". Indeed, Pérez Alfonzo lobbied in favor of "re-dimensioning downwards" the programs in question (Gumersindo Rodríguez, 1979, pp. 513 and 519).

But for Gumersindo Rodríguez these loans would allow Venezuela to maintain a conservationist policy, as the petroleum in the subsoil would appreciate in its value at a higher rate than the interests of the loans. For his part, he qualified the proposition of Pérez Alfonzo as "objectively reactionary" and called it the "most elaborate 'National Destruction Plan' which can be offered by any country" (Gumersindo Rodríguez, 1979, pp. 501 and 513).

In 1930, Juan Vicente Gómez, well aware of what international indebtedness might entail, took the decision to liquidate the public foreign debt in its entirety. During the 1960s, Pérez Jiménez had put the country into debt once again, but this was a floating debt, whose magnitude was manageable. After the fall of Pérez Jiménez, this floating debt was formally converted into public debt. The Betancourt administration chipped away at it with its policy of fiscal austerity and even though levels of indebtedness rose after 1964, until 1973 they remained well below those of the IPR.

Moreover, the concessionaires paid close attention to the public budget, as they were aware that any imbalance would put them in the crosshairs of the fiscal authorities. But with the anticipated reversion underway, they no longer had any reason to care. Hence, a flock of international financial institutions started to descend upon the country, on account of its new credit rating. Public foreign debt grew vertiginously and, by 1978, surpassed the IPR (Figure V.11). The creditors were sure that they would collect on their credits, even if it were on the basis of the future IPR and, even more importantly, that Venezuela would ultimately be forced to re-open the petroleum sector to foreign investment.

Figure V.11 INTERNATIONAL PETROLEUM RENT, EXTRAORDINARY SURPLUS AND PUBLIC DEBT PER INHABITANT 1960 - 2016

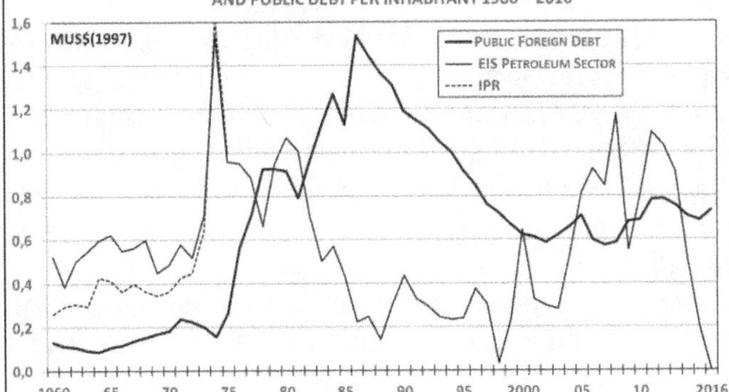

SURCES: FIGURE V.6 AND PERSONAL COMMUNICATION FROM JOSÉ GUERRA.

Pérez Alfonzo, for his part, was acutely conscious that, in the international confrontation over the issue of petroleum, Venezuela would, as an indebted country, be open to all kind of reprisals, about which the country had ample historical experience, furthermore. Given the country's limited capital absorption capacity, there was simply no valid reason to support its becoming indebted.

8. Conclusions

International petroleum rent became increasingly important to Venezuela's economy throughout the whole of the concessionary period. However, its impact on the growth of the non-oil economy was to be a diminishing one to the extent that the economic growth which simply came from its reorganization, based on the international petroleum rent, that is the abandonment of less productive activities in favor of more productive ones, came to an end. Definitively, an increasing part of the IPR would end up by being consumed.

But the Venezuelan landlord state ended up by adopting a position according to which the only way to legitimize the IPR was its investment. Hence, when Betancourt and Pérez Alfonzo had to face a limited capital absorption capacity, they ultimately put forward the proposal of reducing the exports of hydrocarbons. A larger IPR would no longer have any legitimacy whatsoever. Never would Betancourt and Pérez Alfonzo – nor their opponents – consider a higher production and, following the example of Alaska, an investment fund abroad.

Moreover, national public opinion would attribute the failure of President Perez's 'Great Venezuela' to the high propensity of Venezuelans to consume imported goods, to the low intensity of labor and to poor entrepreneurship, in the public as much as in the private sector. Carlos Andrés Pérez himself simply ignored the concept of a limited capital absorption capacity, but so also did Copei's Rafael Caldera, as is clear from his 1975 speech, related to the Nationalization Law Project, when he made reference to the topic of *Petroleum and National Development*. To all of them, the idea of the limited capital absorption capacity of the Venezuelan economy was completely alien.

Carlos Andrés Pérez took office as president at the precise moment when there occurred an explosive increase in the IPR. He decided in favor of a great leap forward. His planning minister blamed the poor academic background of those in charge of national economic policy for the failures of the past, whereas the new generations had been trained in the best universities of the world.

The BCV's National Accounts, for their part, presented the IPR as a national exploitation surplus and, as such, destined to be invested, a characterization questioned by neither the old nor the new generation of economists. Within the framework of these National Accounts, there was no room for the concept of a limited capital absorption capacity. This concept only made sense if one accepts that the IPR was the equivalent of a unilateral international transfer, something which was simply unacceptable to the whole of Venezuela.

Certainly, Pérez Alfonzo and Betancourt were right regarding the limited national capital absorption capacity, but they did nothing to transform this postulate into a *policy*, which, of course, would have implied debating the concept with all the other relevant political forces of the country. To incorporate it in the national political economy would have required, inevitably, that Pérez Alfonzo and, above all, Betancourt, revise their sectarian and exclusive attitude regarding petroleum policy, denying any merit to other political forces and tendencies, as can still be seen, for example, in the discussions about the nationalization law. Pérez Alfonzo and Betancourt entrenched themselves, and never set up a school for preparing petroleum policy maker.

The country, since the 'fifty-fifty split of 1948, was already profoundly divided as far as petroleum policy was concerned. This division is reflected in the positions taken by Carlos Andrés Pérez regarding the Reversion Commission. His speeches reflect the lack of training in the subject matter within AD, which – had it existed – would have allowed him to analyze Venezuela's historic experiences and come to relevant conclusions. Carlos Andrés Pérez, part of a new generation of AD leaders, simply considered that he had the right to a new start, discarding the experience of the past, just as Betancourt and Pérez Alfonzo had done in their turn. President Pérez presented his petroleum and economic policies as radical, eclipsing the policy of *no more concessions*, and nationalizing all the foreign capital that Pérez Alfonzo might have had on his mind.

The national economy showed sustained growth throughout the duration of the concessionary petroleum regime. Between 1925 and 1977, the per capita rate of growth averaged 3.6%, as large investments after 1974 would translate into an immediate increase in the non-petroleum GDP – especially construction – although such investments would never end up to be productive investments. Thereafter the trend was reversed, and the level of the non-petroleum GDP per capita achieved in 1977 would only be surpassed 29 years later, during the years 2006 to 2015.

Part VI.

Venezuela: Cessionary Petroleum Regime

With the entry into force of the Nationalization Law, the major concessionaires became joint-stock companies, fully owned by Petróleos de Venezuela, Sociedad Anónima, the holding company for the operating affiliates, which itself was wholly state owned and the institutional seat of the new cessionary petroleum regime. Creole became Lagoven, Shell de Venezuela became Maraven and the smaller concessionaries agglomerated, during the subsequent years, into Corpoven.

The national petroleum industry as a whole (i.e. PDVSA and its affiliates) became popularly known simply as 'PDVSA', and this is a practice that we shall also follow in this book. However, we will be explicit when we are dealing specifically with PDVSA as a holding company, and as a regulatory entity.

1. Fiscal and Shareholding Control (1976-1988)

1.1. Royalty, Income Tax and Fiscal Export Value

PDVSA's affiliates were subject to the rents and royalties set forth in the 1943 Hydrocarbons Law; however, the higher royalty rates obtained pursuant to the bidding rounds of 1944/5 and 1956/7 were not taken into consideration. The royalty rate, definitively, was one-sixth.

The income tax rate for the affiliates was decreased from 72% in 1975 to 67,7% in 1976. Furthermore, the Nationalization Law provided that this rate applied to only 90% of their export profits, with the remaining 10% accruing tax-free to the holding company.

The Fiscal Export Value (*Valor Fiscal de Exportación* – VFE) was also kept, although it would now simply consist of a percentage surcharge to the export prices determined by CCCH. In this way, the Executive would be able to increase the effective income tax rate in response to changing circumstances in the international petroleum market, without having to go to the National Congress in order to reform the corresponding legislation. Furthermore, the shareholding state, seated in the National Executive, formally would not resort to dividends either, for the same reason.

1.2. Ministry of Energy and Mines and Shareholders' Meeting

The historical structure of the Ministry of Energy and Mines (*Ministerio de Energía y Minas* – MEM) – after nationalization, this was the designation of the erstwhile Ministry of Mines and Hydrocarbons, created in 1951 –, the regulatory entity in charge of the natural resource and of petroleum as a source of international ground rent, was systematically dismantled. In line with this, in 1978, responsibility for the Orinoco Oil Belt was assigned to PDVSA, which proceeded to divide it into three parts: Cerro Negro, Zuata and Hamaca, assigned – that is 'ceded' – in turn to Lagoven, Maraven and Corpoven, respectively.

During the concessionary regime, the Ministry measured daily the volumes produced and determined their quality at the primary storage facilities – within the boundaries of the fields –, free from water and other impurities; royalties due were invoiced on the basis of such measurements. But in 1983, the Ministry turned over the database of the wells to the PDVSA affiliates and, moreover, transferred the personnel qualified to operate it.

Up until nationalization, the Ministry was also in charge of monitoring the issuance of bills of lading at the ports of export. The Ministry determined in its own labs the quality of the exported crudes and petroleum products, and the corresponding volumes were measured with onshore metering devices. With the new cessionary regime, the Ministry transferred these responsibilities to PDVSA's affiliates, which would draft the bills of lading, on board of the receiving tanker and in collaboration with the vessel's captain, on the basis of the vessel's measurements and relying on the quality determined in the laboratories of the affiliates. The Ministry was relegated to the role of putting a stamp on the papers drafted by the affiliates.

Moreover, the Ministry was entrusted with the preparation of PDVSA's annual Shareholders' meeting. But to fulfil this function, the MEM was entirely and exclusively reliant on information and data provided by PDVSA. In consequence, the Shareholders' meeting – presided by the Minister of Energy and Mines and made up of other ministers with responsibilities in the social and economic areas – would be limited to taking cognizance of PDVSA's performance in accordance with whatever the upper management of the company reported. At no point was there ever created a group of professional who would specialize in objectively reviewing and analyzing the performance of PDVSA's variety of national and international businesses.

1.3. Investment Fund, Orinoco Oil Belt and Internationalization

By 1982, PDVSA (the holding company) had accumulated an investment fund deposited in a New York bank in an amount of 5.5 billion US$, which was supposed to have been earmarked for the development of the Orinoco Oil Belt, as a way of putting a brake on the decline of production, and possibly to increase it eventually.

The Orinoco Oil Belt is an immense deposit of extra-heavy crude – i.e. with a gravity lower than 10° API, heavier than water – which is liquid at the original reservoir conditions but becomes a solid at normal surface conditions. In fact, a not dissimilar thing occurred with very heavy crudes which, while less dense than water, were nonetheless quite near to a 10° gravity. In the case of these crudes, the solution to this problem consisted of blending them with lighter crudes, in a proportion such that the resulting blend would have an API gravity of 16° or more, which would mean that it would be a liquid at normal conditions at the surface. This blend could be handled, and in particular exported, like any conventional crude oil.

However, the Orinoco Oil Belt was immense and lighter crudes were scarce. For this reason, PDVSA considered the option of *upgrading* (i.e. partially refining) the extra-heavy crudes to take them to 16° API or more, so that the upgraded crudes could be exported as liquids. As a matter of fact, Lagoven was by then already preparing an association agreement with private foreign investors.

But the demand for OPEC crudes kept decreasing and the country found itself at the door of an exchange rate crisis, as a consequence of its international indebtedness. The government, in an effort to contain the crisis, ordered PDVSA to suspend the corresponding investment projects, and to transfer the fund in question to the BVC, where it would be converted into bolivars. As a consequence, the transferred monies would subsequently devalue, in lockstep with the national currency.

PDVSA responded to this decision of the government with the policy of *Internationalization*. By means of this policy, financial resources would continue to be transferred abroad, but this would be done within a structure that, as a practical matter, would put them beyond the reach of the national government because they would be used to acquire international refining capacities (Boué, 1998). PDVSA, for its part, justified this policy sustaining that it was about securing international markets for the national exports.

In 1983, PDVSA acquired its first shareholding in a refinery abroad, specifically in Western Germany (Veba). Part of the purchase price that PDVSA paid for this acquisition was hidden from its shareholder – and from the country in general – by means of a twenty-year supply agreement, at discounted prices. However, thanks to the existence of the CCCCH, it was relatively easy to detect such discounts. Celestino Armas and Rafael Guevara, both members of AD – then in opposition – estimated the discount conceded by PDVSA at two dollars per barrel.

In 1984, AD won the presidential elections and Internationalization ground to a halt. Nevertheless, given the ongoing shrinkage in demand, the production of OPEC (in general) and Venezuela (in particular) had, by 1985, fallen to half of what it had been in 1973, and prices were on the verge of collapsing. PDVSA would now argue that it needed a free hand in order to compete in the markets for Venezuelan crude, particularly as regards to prices. Indeed, in January 1985, a Joint Resolution of the Ministry of Finance and the Ministry of Energy and Mines provided that, effective from 1984 (i.e. retroactively) the prices to be used for the purposes of assessment of fiscal obligations would be the prices declared by the tax payer; i.e., PDVSA (GOV, 11/01/1985). Forthwith, MEM dissolved the CCCCH and the state thereby gave up, definitively, on auditing the prices of exported hydrocarbons.

Moreover, PDVSA was allowed to reactivate its Internationalization policy. When Royal Dutch-Shell decided in 1986 to shut down its Curaçao refinery – which had been on death row since the 1943 Petroleum Reform – PDVSA leased it in order to keep it open (I. Mommer, 1987). The profitability which

PDVSA would thereafter report for the facility, year in year out, both to its shareholder and to the country in general, would be the result of a transfer price for Venezuelan crude embedding a discount greater than the refining costs.

PDVSA's Internationalization policy would ultimately focus on the US-American market, where it would primarily operate under the name of Citgo. In 1987, 50% of Champlin Refining Company, Corpus Christi, was purchased for 93 million US$ and, at the same time, a twenty-year supply agreement at discounted prices was signed. The following year, the second half of the refinery was bought, this time for 156 million US$. In essence, this price difference reflected the value of the supply agreement with the embedded discounts. However, both to the national government in its capacity as shareholder and to Venezuelan public opinion in general, PDVSA boasted about the good judgment it had shown in purchasing the first half, whose profitability was supposed to be evidenced by the higher price of the second half barely a year later.

1.4. OPEC Quotas and the Orinoco Oil Belt

In 1982, OPEC began considering the necessity to adopt a system of production quotas. This measure would, inevitably, affect the interests of private investors potentially be involved in the development of the Orinoco Oil Belt. But the Venezuelan petroleum minister maintained that OPEC quotas would not be applicable to the Orinoco Oil Belt, as such limits on production were meant to cover only *conventional* crude oil, defined as a liquid both under original reservoir conditions and at standard conditions at the surface. The extra-heavy crude of the Belt was therefore *non-conventional*, as it solidifies at surface conditions. But this argument did not convince OPEC.

By then, the Venezuelan Petroleum Technology Institute was already studying the possibility of mixing two barrels of extra-heavy crudes with one barrel of water, adding an emulsifier to stabilize the mix. Thus, it would be possible to export the extra-heavy crudes as components of a barrel of liquid, with no need to dilute with lighter crudes. Moreover, this Institute had also found that this mixture – which would eventually be called *Orimulsion* – could be used in power plants in the generation of electricity, in the same way as residual fuel oil. Burning Orimulsion, though, would require powerful and costly filters to reduce the emission of pollutants to the atmosphere to tolerable levels.

In 1979, due to the Second Oil Price Shock, the IEA had decided to put an end to the gradual substitution of coal by residual fuel in power plants, and was even thinking of reversing the process. Even though coal was worse in terms of pollution, it was abundant in the consuming countries and it was also much cheaper. This decision would affect Venezuela in particular, given that the country's exports of residual fuels were very significant. Lagoven now took over the project to produce Orimulsion, which it took to calling a '*liquid coal*', albeit coal in the end. As such, it would be competing in the power generation market at *a price competitive with coal*. In 1988, at the initiative of Lagoven, PDVSA created a new affiliate, Bitúmenes del Orinoco (Bitor), which was assigned a block within the Cerro Negro area, so that it could dedicate itself exclusively to the production of Orimulsion[1].

The return on investment for Orimulsion would obviously be far lower. However, PDVSA insisted that the extra-heavy crude processed into Orimulsion would not be subject to OPEC quotas and would therefore never substitute for a barrel of conventional crude, which the country would be able to export as such or else process it into products – both alternatives with much higher returns. Orimulsion would generate a very modest profit, but it would always be an *additional* one.

1.5. Conclusions

With nationalization, PDVSA shook off the fiscal control of the state regarding the production of crude oil and the export of crudes and products in Venezuela's ports, and it also managed to shake off the state control regarding export prices. When it took over the Orinoco Oil Belt, PDVSA maintained that extra-heavy crudes were not conventional crudes and, as such, not subject to OPEC quotas. Moreover, PDVSA was not subject to any shareholder control worthy of the name, as was made evident by – first and foremost – the Internationalization policy.

2. Second Government of Carlos Andrés Pérez (1989-1993)

In 1989 Carlos Andrés Pérez took over the Presidency of the Republic again and Acción Democrática also enjoyed again a majority in the National Congress. This electoral victory reflected a popular hope about better times, fed by the memories of the bonanza of his first presidential period, when his *leitmotiv* had been the 'Great Venezuela'. However, this time Carlos Andrés Pérez announced a 'Great Turnaround'.

At the helm of MEM, he designated Celestino Armas and Rafael Guevara as minister and vice-minister, respectively. Both tried, in vain, to stop the Internationalization policy, but PDVSA was already strong enough to defy even President Pérez. For instance, in 1986, PDVSA acquired 50% of Citgo's Lake Charles refinery for 290 million US$. In 1989, it purchased the remaining 50% for 675 million US$. President Pérez opposed this acquisition; he considered

[1] Lagoven could not sell the same barrel at two radically different prices, in accordance with its subsequent use – i.e. blending it with a lighter crude, or with water to produce Orimulsion – as this would imply an illegal price discrimination. Hence the need to create Bitor, exclusively dedicated to the production of Orimulsion.

it politically unwise for PDVSA to have a shareholding majority in a refinery located in the USA. But PDVSA made its point of view prevail by arguing that an exceptional situation existed, claiming that its partner would sell its block of shares to another buyer that PDVSA considered to be an unsuitable partner. PDVSA therefore acquired the block of shares with the understanding that it would only retain a majority shareholding for as long as it took to find a convenient partner. Two years later, President Pérez personally ordered PDVSA, as a last resort, to sell the second half. PDVSA dragged its feet and never complied, remaining as the sole owner of the refinery.

By 1989, PDVSA no longer saw itself as a mere national state-owned oil company; now it called itself a "Global Energy Corporation" and took to managing its accounts on a global rather than a national basis. In consequence, to the direct acquisition costs of its refining and storage capacities abroad would thereafter also be added all related costs.

2.1. International Monetary Fund and the Caracazo

In his second period in office, Carlos Andrés Pérez had to face a fiscal situation which was radically different from the one he had encountered during his first presidential period: the coffers were empty and the state indebted. He was thus forced to resort to the International Monetary Fund to request an emergency loan for five billion dollars. The International Monetary Fund would not only grant this loan on condition that the national economy be opened to foreign investment, but also to increase the prices of hydrocarbon derivatives in the domestic market. On February 26, 1989, the government increased domestic prices by 94%, on average, and made a commitment – by signing a Letter of Intent with the International Monetary Fund on February 28 – to raise them further by 70% by January 1990, and thereafter by half-yearly adjustments throughout the subsequent two years, until prices in the domestic market reached international levels.

However, the increase of February 26th had the immediate consequence of a rise in the price of public transportation. The next day, on February 27th, this triggered a popular rebellion known as the "Caracazo". Carlos Andrés Pérez and Acción Democrática – the erstwhile party of the masses – would have no answer other than resorting to military force, with a casualty toll counted in the hundreds. This event sounded the death knell for the political regime then in force.

2.2. Oil Opening

The policy of opening up to private foreign investment would soon reveal itself as an Oil Opening (*Apertura Petrolera*). In September 1990, the Minister of Energy and Mines, Celestino Armas, sent the President of PDVSA, Andrés Sosa Pietri, the instruction to open up a space for private investment through Operative Agreements (OAs). These, per article 5 of the Nationalization Law, were not important, since they were supposed not to 'affect the substantial nature of the activities attributed' and were therefore left to PDVSA's sole discretion, without there being any need of obtaining the authorization of the National Congress. Now, the Minister informed PDVSA of the government's decision that:

> "The operating affiliates [...] are to reactivate those fields which they have abandoned in their assigned areas and which still contain volumes of hydrocarbons in their deposits. In order to fulfill this objective, the operating affiliates can sign the [OAs] which they consider necessary with private enterprises interested in such reactivation. The operating affiliates shall maintain total control over any hydrocarbons produced pursuant to the aforementioned agreements" (Armas, 21/09/1990; quotation marks in original).

Even before this notification, Lagoven was already well advanced in drawing up an association project denominated Cristóbal Colón. This project was to consist of the exploration and exploitation of non-associated natural gas deposits – discovered recently in the Gulf of Paria – and its subsequent liquefaction and export. Lagoven agreed with Exxon and Shell, its potential partners in the venture, that the former would first obtain a "pronouncement from the Supreme Court of Justice with regard to the supremacy of the Nationalization Law of 1975" (Lagoven, 03/1993, p. F-1).

2.3. Request of Clarification before the Supreme Court of Justice

Indeed, in November 1990, Lagoven put before the Supreme Court of Justice a *Request for Clarification*. In particular, it asked for the derogation of art. 3 of the 1967 Hydrocarbons Law, on grounds of incompatibility with article 5 of the Nationalization Law. As noted before, with the approval of article 3 of the 1967 Hydrocarbons Law, a space had been opened within the concessionary regime for CVP's Service Contracts; but the concessionary regime by this time had been abolished by the Nationalization Law. Now, the Court sided with the arguments presented by Lagoven and held that the intention of the legislature in the Nationalization Law had been to regulate the exploitation of hydrocarbons "*ex novo*, so therefore any preceding general norm referred to this subject matter should be considered derogated". This conclusion was based on the last article in the Nationalization Law, which provided that "the dispositions of the Hydrocarbons Law [...], and any other, *which might collide with the present law* are hereby derogated" (Supreme Court of Justice, 23/04/1991; emphasis ours). Given that the statutes at issue were an ordinary law and a subsequent organic law, the Supreme Court of Justice concluded that any pre-existing legal position based on the concessionary regime had been repealed. The Court therefore held article 3 of the 1967 Hydrocarbons Law to be derogated (Vallenilla, 1995).

As this article was derogated in whole, this meant that its provisions stipulating that the enterprises that were wholly owned by the state and the agreements promoted by them were to be subject to exclusive national jurisdiction and tax sovereignty, no longer applied.

The Judgment was unanimous. The request for clarification, after all, had been made by Lagoven, affiliate of the state oil company, and not by Exxon or Shell. No one else made an appearance before the Court, not even a representative of the Ministry of Energy and Mines, because it was assumed that this request was aimed merely at clarifying the applicable legal regime relevant only to internal procedures regarding the Cristóbal Colón Project, and was devoid of any wider significance.

The judge who wrote the decision, Román J. Duque Corredor, happened to have been Legal Counsel for Maraven from its foundation and up until 1979. In 1978, Duque Corredor had published a book entitled The Law of the Nationalization of Petroleum (*El Derecho de la Nacionalización Petrolera*), in which he made reference – in its first line, to be precise – to the "repealed Hydrocarbons Law", and further clarified the matter in a footnote stating "repealed by the [Nationalization Law]".

2.4. Petróleos de Venezuela, S.A.

Carlos Andrés Pérez never appointed a career oilman to the presidency of PDVSA. The role of the President of the state enterprise and of its board of directors was to supervise the operating affiliates and protect them from any political interference. Conversely, the inevitable and necessary political interaction of the affiliates with the government or with the country at large had to go through the intermediation of the board of directors of the *holding* company, PDVSA. When PDVSA first came into being, Pérez designated as its President General Rafael Alfonso Ravard, who had had a long and distinguished career in the public enterprise sector. In his second period in office – in March 1990 – Carlos Andrés Pérez designated Andrés Sosa Pietri, an industrialist and ex-senator as president of PDVSA. Pérez made this designation against the explicit wishes of the PDVSA upper management, which considered that this post should always be occupied by one of its members.

Sosa Pietri was a straight talker, an absolutely exceptional characteristic in the PDVSA milieu. He became the President of PDVSA at the precise moment when the Oil Opening was about to begin. In July 1990, he delivered a conference to the IV Venezuelan Petroleum Congress entitled "The Response of the Petroleum Industry to the Venezuela of Today". In his words, PDVSA would have to put to the government:

> [...] a change in our negotiation strategy within [OPEC] [...], which has been progressively weakening, especially after the establishment of the quota system in 1982, and to reassert our legitimate rights [...] for a fairer deal in accordance with its condition of being the sixth largest holder of conventional petroleum reserves (and the first when [...] the Orinoco Belt is added) (Sosa Pietri, 18/07/1990, p. 5).

In 1982, when OPEC quotas were introduced, the participation of this organization in the world petroleum market was above 60% and the price of the Arabian Light marker crude was at 28 US$/b. Eight years later, OPEC's market share was less than 36% and the reference price had fallen below 17 US$/b. According to Sosa Pietri, this was a consequence of the adoption of the OPEC quotas, which he saw as representing "a totally wrong strategy". Even worse, in 1970, Venezuelan production represented one sixth of OPEC output, but less than one twelfth by 1990. Moreover, he considered that, "thanks to its 'Internationalization' policy, Venezuela currently has more market outlets than those it can satisfy with its OPEC quota. This forces us to buy crude from other sources in order to cover them".

Sosa Pietri proposed to produce at full blast to reach 3.1 million b/d by 1995, ignoring OPEC quotas (Sosa Pietri, 1990, p. 10). However, not only did he run into a definite 'no' from Minister Armas but also from President Pérez. Nevertheless, in his book Oil and Power (*Petróleo y Poder*) – which he wrote immediately upon leaving the presidency of PDVSA in March 1992 – he would continue to insist on:

> [...] the obvious conclusion [...] that Venezuela has to redefine its role within OPEC, contribute to a substantive reform of this organization and the role it plays, or else separate itself completely from it (as Ecuador has just done) (Sosa Pietri, 1993, p. 86).

Sosa Pietri held, without a shade of doubt, that OPEC had always been a failure and, according to him, this was proven by the fact that oil prices in the world market had continued to slide throughout the 1960s, and the subsequent increases registered during the 1970s he saw as entirely accidental. "During its first thirteen years, OPEC had not managed to influence oil prices in any way, and even less to anticipate the great increases of 1973" (Sosa Pietri, 1993, p. 58). All in all, Sosa Pietri considered that "OPEC is a myth, and Rómulo Betancourt and Juan Pablo Pérez Alfonzo are legends" (Sosa Pietri, 1993, p. 90).

Conversely, PDVSA should:

> [...] behave with the same liberty as any commercial company; as if it were, to all legal purposes, a "private company" and not a sector of the state. In this way was born a creature which, in the not too distant future, will escape its creators and, what is more, will bury both the petroleum policy of old and the stellar role played by the Venezuelan political sector in the country's petroleum activities (Sosa Pietri, 1993, p. 63; quotation marks in the original).

For Sosa Pietri, the responsibility of the Minister of Energy and Mines in PDVSA's shareholders' meeting should be limited to being "the moderator of its sessions" (Sosa Pietri, 1993, p. 65). Unfortunately, when Sosa Pietri took office as President of PDVSA, he found:

> […] that the managerial and financial autonomy […] have been fundamentally violated. The investments and operations budgets are no longer the exclusive competence of PDVSA, but of its shareholders' meeting; that is to say, of the Ministry of Energy and Mines. All plans and programs, […] are subject to the 'guidelines' […] of the 'Minister'. […] The 'Minister' regulates oil production, using OPEC, and actively participates in the mechanism for setting prices, which are not considered to be definitive until he expressly signs off on them (Sosa Pietri, 1993, p. 77f).

In fact, Sosa Pietri designed an ambitious plan to transform PDVSA into a great global energy corporation, but both Minister Armas and President Pérez ordered him to abide by OPEC quotas.

In August 1990 Iraq invaded Kuwait, and the quota system was temporarily defused. Anyway, PDVSA did not have the funds at hand to finance its expansion plan. Sosa Pietri then proposed to turn PDVSA into a truly private corporation "through the placement of the company's shares in certain national and international stock exchanges (especially the latter)". Thus, the state would retain a shareholding majority albeit in a way in which "while not forfeiting control of the enterprise, actually restores its autonomies to the latter, and obliges the state to apply to it a rate of taxation which is in accordance with that of an industrial business to which private investors are to be attracted" (Sosa Pietri, 1993, p. 79).

Sosa Pietri had no doubts that any intervention by the state – that is, after all, the *shareholder state* – which could restrict PDVSA's competitiveness, was counterproductive. He attributed both indebtedness and the failure of the Great Venezuela to the incompetence of those in government and, of course, to corruption. In his view, it was necessary to save PDVSA from its sole shareholder, that is, the state, in order to prevent the company from being buried under the rubble of the economic collapse of the country.

Andrés Sosa Pietri called a spade a spade. Nevertheless, his position only reflected the dominant and widespread – even if *sotto voce* – discourse within PDVSA and the national business sector in general. Petroleum would already only matter as a productive activity. For sure, with nationalization, the country had grown more divided than ever before.

2.4.1. Orimulsion

Sosa Pietri also developed an enthusiasm for Orimulsion:

> It was about a new fuel […], a kind of liquid coal, but with the thermal properties of the best quality coals. There was a market […] big enough to ensure, in competition with coal, the placement of around 1.5 million barrels per day of crude with less than 10° API.
>
> Once I became familiar with the project and its advantages, I gave it my unequivocal and enthusiastic support (Sosa Pietri, 1993, p. 121f).

The position taken by PDVSA, that the corresponding production of extra-heavy crude would not be subject to OPEC quotas, was backed by the International Energy Agency. The latter, in January 1993, stated in its Monthly Oil Market Report that it would no longer include such crudes as part of Venezuela's crude oil output, and would instead place them in the category of 'non-conventional oil' (Boué, 2012).

2.5. Venezuela, Coup by Coup

In February 1992 there occurred a first attempt at a military coup, headed by Lieutenant Colonel Hugo Chávez, motivated by the failure of the economic policies of the previous fifteen years, and for which Carlos Andrés Pérez was undeniably responsible. This first attempt failed, as did a second one in November 1992. From within his prison cell, Hugo Chávez also claimed the leadership of the latter.

With the first coup attempt, in any event, the government of President Pérez had been weakened, which led it to embrace an alliance with Copei. He was thus forced to substitute Minister of Energy and Mines, Celestino Armas, with Alirio Parra, who had been a member of PDVSA's first board of directors.

Within this already complex situation there occurred a third coup, one which had a very different origin, being legal-political in nature. This last coup put an end to Pérez's presidency in May 1993, barely eight months before his constitutional period in office was supposed to end. Amidst a severe economic crisis and major political turmoil, historian Ramón J. Velázquez – independent, albeit close to Acción Democrática – took office as Interim President. The fall of Carlos Andrés Pérez marked a dividing line in Venezuelan petroleum policy. From that point on, it would definitely be in the hands of PDVSA, who was to impose it against the Venezuelan state, that is to say, against its sole shareholder.

Now, in November 1993, as Bitor was about to sign its first Orimulsion supply agreements with foreign electricity generation plants, the Ministry signed a royalty agreement with Bitor pursuant to which the royalty liquidation price was reduced by approximately 30% in comparison to that applicable to the same crude when it was mixed with lighter crudes (MEM, 11/11/1993).

3. Operative Agreements

In 1991, in fulfilment of the instructions from Minister Celestino Armas, PDVSA organized a First Bidding Round for Operative Agreements for abandoned marginal fields. In 1992, Minister Alirio Parra addressed the National Congress to report on the results of the round and to ratify that, indeed, these were OAs, and hence were of the sole competence of PDVSA and, therefore, did not require to be approved by the National Congress.

According to the Minister, the model contract used in this First Round was not an Association Agreement (AA), because it would "not give the contractor any degree of participation or association with the contracting PPDVSA affiliate" (Parra, 04/08/1992). For its part, PDVSA put before the Permanent Commission for Energy and Mines of the Chamber of Deputies the expert opinions of renowned lawyers, all of whom agreed that the model agreement used fulfilled the requirements for an OA set out in art. 5 of the Nationalization Law. They were not AAs, as there was no sharing of profits or losses by the affiliates, and contractors would only be paid a fixed amount per barrel as an 'incentive', whose magnitude would be determined through bidding. That said, all of these experts consulted had been retained and put forward by PDVSA and its affiliates and, obviously, they were not going to present dissident expert opinions, which were therefore condemned to silence. Finally, the Permanent Commission reported to Congress that the contracts involved were, indeed, OAs falling under the sole competence of PDVSA, a posture which the legislative organ accepted without further discussion.

As a matter of fact, however, the contractors were producers of petroleum, and the affiliates had no presence whatsoever in the fields. Yet for the time being, these contracts were of little practical importance as the fields involved were marginal and abandoned. But the concept of an OA would not be subject to scrutiny thereafter, even though PDVSA would no longer use the same model contract. In the First Round, the abandoned marginal fields were defined on three dimensions, that is to say, not only in terms of their surface but also their depth, and contractors were not allowed to drill into deeper unexplored strata. In the Second Round, fields would only be defined in two dimensions, that is, in terms of their surface coordinates. Then, in 1995, Maraven negotiated an OA with Chevron for the Boscán field, which was producing 80 thousand b/d at the time; it was definitely not a marginal field. Finally, in 1997, in the Third Round, the fields put out to bid were once again marginal but they were in production, to the tune of 70 thousand b/d and, additionally, new contiguous areas for exploration and subsequent exploitation were also included.

With the OAs, production activities were definitively privatized.

3.1. International Commercial Arbitration: Operative Agreements

Furthermore, the conditions for the First Bidding Round for operative agreements established that any controversy that might arise between the affiliates and the contractors was to be definitively submitted to *national commercial arbitration* which would take place in Caracas, in accordance with Venezuelan law. Now, according to article 3 of the 1967 Hydrocarbons Law, any agreement by an enterprise wholly owned by the state had to include, obligatorily, the following clause:

> "Doubts and controversies of any nature which may arise under this agreement and which cannot be resolved amicably shall be decided by the competent

Tribunals of Venezuela, in accordance with its laws, and for no motive or cause shall they be the origin of foreign claims" (Quotation marks in the original).

However, the Supreme Court of Justice had recently invalidated this article.

For OAs assigned during the Second Round, in 1992, arbitration was still going to take place in Caracas, but it would now be subject to the rules of the Paris-based International Chamber of Commerce (ICC). This institution, founded in 1919, is the most important forum in the world for international arbitrations of commercial disputes between private parties; and PDVSA and its affiliates were *private* companies.

Finally, in 1997, conditions for the Third Round provided for arbitration to take place in New York. From then on, dispute resolution would definitively take the form of a full-fledged international commercial arbitration.

3.2. Fiscal Regime

The duration of these agreements was set at twenty years even though, due to their status as purported OAs, PDVSA was entitled to extend and/or modify them at will. It was the responsibility of its subsidiaries to approve the annual program of activities, as well as the corresponding budget, and all of the investment destined for the area in question was to become the property of the affiliates albeit in the form of a *contingent liability*. In the First Round OAs, the contractors were to recoup their operational costs in the first place, and the contingent liability in the second place. But by the Third Round, investments were to be given the same treatment as operational costs for accounting purposes.

Throughout the concessionary regime, the petroleum sector had been ring-fenced with regard to income tax, to prevent the concessionaires from transferring windfall profits to other sectors where lower tax rates prevailed. For this reason, any service enterprise which worked for the concessionaires was likewise subject to the petroleum sector tax rate. With nationalization, this provision disappeared, which in turn meant that when the OAs, by putting themselves forward as mere service enterprises before the country's tax collection agency – the Integrated International Customs and Tax Administration Service (*Servicio Nacional Integrado de Administración Aduanera y Tributaria* – SENIAT) – would be subject to the non-oil income tax rate of 34%, and not to the petroleum rate of 67.7%, as would have been the case under the concessionary regime.

Royalties were to be paid by the affiliates. For First Round OAs, the government reduced the applicable rate to 1%, since the fields involved had been abandoned. For the Second Round OAs and the Boscán OA, the royalty rate was one sixth. However, for the Third Round OAs, the contracts in question only made reference to the 'legal rate'. At that point, PDVSA was already considering a legal reform whereby the customary royalty rate would be reduced from 16⅔%, to 5% (PDVSA, 15/03/1997, p. 4).

3.3. Conclusions

The fiscal regime of the OAs followed the same trend, with regard to the reduction of royalties and income tax rates, that we observed in connection with the *Petroleum Revenue Tax* in Great Britain. Any part of the *contingent liability* that had not been recovered at the end of the agreement would have to be absorbed, *for the time being*, by the contractor as a loss. But there already was a precedent set by the Boscán OA, where PDVSA had bound itself to paying off any pending part of the *contingent liability* in whole. Furthermore, there was nothing to prevent the affiliates and their contractors from modifying the terms of the OAs in the future.

By 2005 there were 32 operative agreements in force, covering a surface of 44.7 thousand km², and producing approximately 500 thousand b/d. Throughout the whole period between 1993 and 2005, payments to the contractors came to 55% of the export price (with all profits subject to the non-oil income tax rate of 34%). The remaining 45% corresponded to PDVSA's affiliates, which had to pay the corresponding royalties as well as a petroleum sector income tax rate of 67.7% (Ministry of Energy and Petroleum, 08/2005).

4. Association Agreements

In June 1991, the Minister of Energy and Mines, Celestino Armas, presented the Cristobal Colón Association Project to the National Congress. In fact, it was only a *Preliminary Development Accord* signed by the partners in February 1991, whose purpose was to evaluate the feasibility of the Project. The Minister at that time sent a communication that ended with the following paragraph:

> In order duly to inform the Congress of the Republic of all the pertinent circumstances to the association agreement and with the best disposition to provide any additional information that may be required of us, we are submitting the annexes […], among which is included a proposal of conditions for the execution of said agreement, structured on the basis of conversations that Lagoven, S.A., an affiliate of Petróleos de Venezuela, S.A., has been holding with the potential associates, elements which, as a whole, we respectfully submit for the consideration of this sovereign Congress, with the aim of complying with the requirements of art. 5 of the [Nationalization Law] (Armas, 06/14/1991).

Annex C referred to the *Proposed Conditions for the Association Agreement to be considered by the Congress of the Republic*. What government wanted from the National Congress was that this Framework of Conditions (FOC), agreed between Lagoven, on the one hand, and Exxon and Royal Dutch-Shell, on the other hand, be confirmed and then subsequently published in the Official Gazette, as the FOC set by the 'Chambers in joint session'.

4.1. International Commercial Arbitration: Framework of Conditions

The proposed FOC stressed the need for international commercial arbitration:

> Any controversy or claim that may arise in connection with the [AA], the execution of which is hereby authorized, shall be finally and definitively settled by international arbitration in accordance with the rules of the [ICC], in the city of New York (Condition 20).

But now, it was no longer simply a matter of international commercial arbitration, but also of *limiting the sovereign rights of the state*:

> The [AA] to be signed shall include provisions that would allow Lagoven to compensate, on equitable terms, the foreign shareholders for the adverse patrimonial consequences [...] derived [...] from the adoption of arbitrary decisions of national, state or municipal administrative authorities or changes in legislation that, due to their content and purpose, were to lead to a discriminatory treatment against the Company or said shareholders [...] (Condition 13).

Furthermore, in the event of an arbitration award in favor of the private associates, they latter would not have to resort to the national Courts for its enforcement. Although Lagoven did not have assets abroad, PDVSA certainly did thanks to its Internationalization policy, and the Framework of Conditions established that:

> If the organizations Shell, Exxon and Mitsubishi, in order to sign the [AA] [...] act through subsidiaries, then they shall grant ample and sufficient surety, as required by law, to guarantee all the obligations acquired or arising for said subsidiaries from the referred [AA] and [...] from its execution. PDVSA shall do the same in relation to its subsidiary Lagoven, in reciprocal terms (Condition 15).

This 'reciprocity' was, of course, misleading. Lagoven gave guarantees against certain actions of the Venezuelan state, whereas Shell, Exxon and Mitsubishi gave no guarantee against any action by The Netherlands, Great Brittan, USA or Japan. Finally, in the proposed FOC, it was stated:

> The [AA] to be entered into, the commercial company to be created and the various activities that will derive from such acts, especially the commercial ones, are operations and businesses that only concern and bind Lagoven or [PDVSA], to the extent of the guarantees that it may grant in accordance with Condition 15 and, in no case, engage the responsibility of the Republic of Venezuela (Condition 18).

In other words, although the jurisdictional and tax sovereignty of the Venezuelan state was not going to be formally limited, the National Congress was going to have to accept that PDVSA would instead do so in the AAs. Hence, the *sovereign rights of the state were subject to international commercial arbitration*, through the interposition of PDVSA, created under the law of *private* corporations, but which was 100% state-owned.

Hence, the Internationalization policy reveals itself as a preliminary phase of the Oil Opening, and it was not only about the capital invested abroad by PDVSA, but also about crude stored abroad and the possibility of attachment of the sizable accounts receivable of PDVSA's international subsidiaries.

Then, when the FOC was approved, the preamble of the corresponding publication in the Official Gazette stated that the National Congress:

> After studying and discussing the Report presented by the Commission in a joint session of the Legislative Chambers [...], agreed [...] to formally authorize the Association [...].
>
> This authorization must be used within the legal framework of the "conditions" expressed specifically in said Report, which fully guarantee total compliance with the conditions of legality, legitimacy, opportunity and convenience, expressed in the sole paragraph of Article 5 of the [Nationalization Law] (GOV, 09/09/1993; emphasis ours).

In reality, it was not up to the Legislative Power to pass judgment on the 'full compliance' of the FOC with article 5 of the Nationalization Law, as this was an attribution of the Judiciary; in the same way that it is not up to the Legislative Power to pass judgment on the constitutionality of a law. However, in any international commercial arbitration where PDVSA or one of its subsidiaries would feature as respondent, it would be practically impossible for them to mount a defense centered on questioning the FOCs, given that the National Congress itself had expressly and explicitly recognized it to be in accordance with article 5 of the 1974 Nationalization Law.

4.2. International Commercial Arbitration: Association Agreement

In that same Annex C of the Cristóbal Colón Project, Lagoven stated that:

> The [AA] must be submitted by the National Executive to the Congress of the Republic for its authorization (Lagoven, 06/1991, Annex C, p. 14).

Later on, upon reactivation of the Cristóbal Colón Project in March 1993, Lagoven sent new documentation directly to the Commission, including a draft of the *Association Agreement and its Annexes*, as well as a *Formation Accord* that the partners had just signed. The *Formation Accord* now stated that "the Congress has demanded a copy of the draft of the said Agreement" and that:

> Immediately after the signing of this Accord, Lagoven shall make the necessary arrangements for this Accord and the full text of the [AA] to be submitted to Congress, in connection with the aforementioned request for authorization of June 14, 1991 […] (Lagoven, 03/1993, Annex IV, p. 1).

However, Lagoven and its partners dragged their feet on the Cristóbal Colón Project and, in fact, never drafted such an AA. Nevertheless, in 1995, when PDVSA organized a bidding round for ten Association Projects for the *Exploration at Risk of New Areas and the Production of Hydrocarbons under the Profit Sharing Scheme* (Profit Sharing Projects), the National Congress approved the corresponding FOC, but now, in the Official Gazette, it stated that the National Executive:

> […] submitted on December 7, 1994, a Draft Framework of Conditions for the conclusion of […] Profit Sharing [AAs]; […]; that the Bicameral Commission of Energy and Mines has carried out a detailed analysis of the framework of conditions that has been proposed for the [AAs];
>
> that according to Article 5 of the [Nationalization Law], the authorization of the National Legislative Chambers meeting in joint session is required, *both for the Framework of Conditions and for the texts of the association agreements themselves* […] (GOV, 17/07/1995; emphasis ours).

What was being authorized was not an Association Project as such, but only a 'Draft Framework of Conditions', which moreover would have been the object of a detailed analysis by the Commission. However, according to PDVSA's interpretation of article 5 of the Nationalization Law, the National Congress still needed to authorize the AAs themselves. This article gave the National Executive and PDVSA the faculty to enter into AAs:

> […] with a participation therein as will guarantee control by the state, and establishing a fixed duration. Authorization from both Chambers of Congress meeting in joint session shall be required prior to entering into such agreements, under such terms as may be set therefor […].

Obviously, the above conditions should have been part of the FOC. But now the National Congress accepted Lagoven's interpretation that the same AAs had also to be authorized.

Indeed, these AAs were duly submitted to the President of the Republic in Council of Ministers, and after having been granted the corresponding approval – within a period of time that could not extend more than eight days – the MEM sent them to the National Congress. The Bicameral Commission confirmed that the AAs complied with the FOC, and the National Congress in turn approved them, based on the consideration that the Bicameral Commission would have "carried out a careful analysis of the content" of the agreements in question, and would have likewise verified that they would be "in accordance with the [FOC] established by the Congress of the Republic" (GOV, 06/26/1996).

In reality, this was not exactly what happened. The drafting of the FOC and, therefore, of the AA, involved teams of professionals drawn from the most diverse specialties, involving not only technical and economic aspects, but also Venezuelan law, the national law of the associates, and international law. Moreover, they were drafted in a way meant to confuse deliberately any reader in the Venezuelan political world, and also with a view to their subsequent re-interpretation. In reality, neither the National Executive, nor the Bicameral Commission, nor the National Congress, whether acting solely or jointly, could have understood them to their full extent, let alone comprehended their implications. Indeed, the competence of the National Congress was a *political* one, as had been evident in its involvement with respect to the service contracts of yore, but now PDVSA and its foreign associates imposed their viewpoints and turned the *Oil Opening* into a policy of *experts*, and all the experts were going to work with or for PDVSA. Thus, in any international arbitration – whenever PDVSA or one of its subsidiaries were to be sued – it would be practically impossible for them to challenge, in their defense, any clause of an AA, since the Executive and the National Congress would have approved it, the most arbitrary and aberrant interpretations of the FOC included.

4.3. Arbitrated Sovereignty

For the time being, the guarantees in question could only come from PDVSA. However, the ground was being prepared for them to be assumed by the state itself. Indeed, in October 1991, the Venezuelan government signed a first bilateral investment treaty (BIT), the Agreement on Encouragement and Reciprocal Protection of Investments between the Kingdom of the Netherlands and the Republic of Venezuela (*Convenio para el Estímulo y Protección Recíproca de las Inversiones entre la República de Venezuela y el Reino Unido de los Países Bajos*), with a term of fifteen years, automatically extendable (GOV, 06/08/1993). This treaty granted Dutch investors in Venezuela and Venezuelan investors in The Netherlands, the irrevocable right to file for international arbitration against the respective state. It would not be the Dutch and Venezuelan investors who would request international arbitration from the Venezuelan and Dutch governments but, instead, it would be both governments who would offer it, unilaterally and unconditionally. The foreign investors, if necessary, would only have to send a letter to the respective government accepting the offer to arbitrate, in the event of a dispute arising. The investors, for their part, were not consenting to arbitration against themselves, as they were not parties to the treaty. Thus, the states would always be the defendants, and the investors the claimants. The Convention contained a list of the "Investments" covered, among them:

> [...] rights granted under public law, including rights to prospect, explore, extract, and exploit natural resources.

However, in 1991, when the BIT with the Netherlands was first presented to the National Congress, there was much resistance against it. Nevertheless, Lagoven and its partners in the Cristóbal Colón Project insisted that it be duly approved by the National Congress, ratified by the President of the Republic and published in the Venezuelan Official Gazette. As long as this did not happen, they were not going to go ahead with the Project. In fact, Royal Dutch-Shell – a Dutch company par excellence – had participated in the First Round of OAs in 1991, and had won the bidding for the Pedernales field. However, Shell relinquished the OA in question because, according to the *Petroleum Economist*:

> An axiom of Shell's foreign investment policy is that there should always be a guarantee of international arbitration, because it has no confidence in a host country's legal system (Kielmas, 11/1992).

In the Second Round of OAs, Shell participated again, and this time won the Urdaneta Oeste field. Shell signed the corresponding OA in November 1993, *after* the BIT with Holland was approved by the National Congress, ratified by the President of the Republic and published in the Venezuelan Official Gazette.

The most elementary of all sovereign acts, involving the national territory and its constituent parts, was thus subject to international arbitration. However, article 4 of the 1943 Hydrocarbons Law – which was kept without change in the 1967 Hydrocarbons Law – had provided that any concession title should, mandatorily, include the following clause:

> "Any doubts and controversies of whatever nature that may ensue because of this concession and which cannot be amicably settled, shall be decided upon by the competent courts of Venezuela, and in accordance with its laws, and for no reason nor for any cause shall they give rise to foreign claims" (Quotation marks in the original).

But the Nationalization Law, in its first article, had put an end to the concessionary regime.

Regarding the bilateral nature of this treaty, it has to be clarified that both Dutch investors in Venezuela, as well as Venezuelan investors in the Netherlands, are defined as such by the respective national legislation. According to Dutch law, all that is required to qualify as such is no more than a post office box in Holland and the hiring of the services of a lawyer to ensure compliance with the minimum legal requirements, all without having a single employee in the country. Thus, in Venezuela, not only Royal Dutch-Shell, but also the Italian Ente Nazionale de Idrocarburi, the US-companies ExxonMobil,

Chevron and ConocoPhillips, the Chinese CNPC and the Norwegian Statoil were going to present themselves as Dutch investors, when it suited them.

According to the Venezuela-Netherlands BIT, the arbitral forum was to be the *International Centre for Settlement of Investment Disputes* (ICSID), an agency of the World Bank based in Washington. Created in 1966, pursuant to the *Convention on the Settlement of Investment Disputes between States and Nationals of Other States*, the countries that became members undertook to recognize ICSID awards as *final and definitive decisions*, which would no longer require any confirmation for enforcement within their territories, not even from the national Supreme Courts. Venezuela was still not a member of ICSID, but this did not prevent an investor from resorting to ICSID arbitration against Venezuela, provided that the claimant qualified as an investor of a country that was a member, as was the case of the Netherlands. In all, between 1993 and 2008, Venezuela signed, the National Congress approved, and the Presidency ratified twenty-eight similar bilateral investment treaties, all of which were then duly published in the Official Gazette.

In June 1993, the government of Ramón J. Velázquez signed the Venezuela-Netherlands BIT, the National Congress approved it in 1994, and President Rafael Caldera ratified it in 1995.

4.4. Relevant Circumstances, Agreed Conditions and Fixed Conditions

Between 1991 and 1997, the National Executive presented seven association projects to the consideration of the National Congress (Table VI.1).

Table VI.1		
ASSOCIATION PROJECTS 1991–1993		
SUBSIDIARY	PROJECT	CONGRESS
Lagoven	Cristóbal Colón Liquefied Natural Gas	1991–1993
Maraven	Petrozuata Upgrading	1993
Maraven	Sincor Upgrading	1993
Subsidiaries	Profit Sharing	1994–1996
Bitor	Florida Power & Light Orimulsion	1996
Corpoven	Hamaca Upgrading	1996–1997
Lagoven	Cerro Negro Upgrading	1996–1997

In accordance with the Judgement of the Supreme Court at the beginning of 1991, the authorization of the AAs would have as its basis, solely and exclusively, the second paragraph of article 5 of the Nationalization Law. This authorized PDVSA and its subsidiaries:

> In special cases, and whenever it so suits the public interest, [...] to enter into association agreements with private entities maintaining such a participation therein that will guarantee control by the state, and establishing a fixed duration.

Let us see how the interpretation of this second part of article 5 of the Nationalization Law developed throughout this process.

4.4.1. State control

According to the Judgement of the Supreme Court just quoted, the participation of the state as a majority shareholder was not necessary in order to comply with the "control by the state" required by article 5 of the Nationalization Law. Moreover, already in 1991, among the Conditions Agreed by PDVSA/Lagoven and its partners with respect to the Cristóbal Colón AA, in Condition 9, a minority shareholder participation of Lagoven in the association was stipulated as obligatory. This condition was to be present in all the AAs.

Consequently, control by the state was to be defined by some legal arrangement, and was to be exercised through the participation of the subsidiaries in the AAs. The kind of decisions contemplated in this arrangement were divided into three categories. In the first, unanimity would be required, so that affiliates would have a right of veto, as would all other partners. The second category would require a qualified majority, which was defined in such a way that the affiliate would still have a right of veto, although to impose a decision it would also need the supporting vote of some associate. Finally, the third category would only require the majority of the voting shares and, therefore, would not need the positive vote of Lagoven as a minority partner. In this category would be included all decisions concerning the association's activities related to production. These three categories are listed as such within the FOC, although their definition was left to the discretion of the associates. Subsequently, when PDVSA insisted, ultimately successfully, that even the AAs be approved by the Executive and the National Congress, these definitions ended up by being legalized.

The control structure was to change in subsequent associations, although its essence would remain the same. Thus, for the Profit Sharing Projects, in accordance with Condition 4 of the FOC, a "Control Committee" was to be created, independently from any shareholding. The Control Committee would be "formed by an equal number of members appointed by the Investors and the Affiliate, which will be chaired by a member appointed by the latter", "with the Chairman having a casting vote to resolve tied cases". The Affiliate in question was finally to be the CVP, which PDVSA had reactivated as an affiliate for this specific purpose. In particular, Condition 4 stipulated that the associates were to:

> […] submit to the Control Committee for approval key decisions of national interest related to the implementation of the Agreement (GOV, 17/07/1995).

Thereafter, in the AAs, what this 'national interest' would consist of was specified in the following terms:

> The members of the Control Committee appointed by the CVP shall vote on any matter or issue submitted to the Committee […], taking into account whether or not the proposal in question is consistent with the national interest of the Venezuelan state in the exploration, development and exploitation of its hydrocarbon reserves, *including the national interest of the Venezuelan state in attracting and preserving private investment projects of importance to the national economy.*

Even more:

> In the event that CVP's representatives vote against a proposal that is favorably voted on by the representatives of the Investors, they shall express in writing the reason for voting against the proposal, identifying the manner in which such proposal is inconsistent with the national interest of the Venezuelan state. In this regard, *the Investors shall have the right to undertake any applicable actions before the competent Venezuelan authorities* (Corpoven, 1995, Article 4.7; emphasis ours).

4.4.2. Fixed term

The association agreements, according to article 5 of the Nationalization Law, had to have a 'fixed duration'. The FOC proposed by Lagoven for the Cristobal Colón Project, in 1991, specified that:

> […] the partners shall state, by 12/31/96 at the latest, their irrevocable decision to invest […]. This term may be extended for up to two years at Lagoven's discretion (Lagoven 06/1991, Condition 10).

In 1993, in the new documentation sent directly to the National Congress, Lagoven replaced the last sentence with the following:

> If such decision has not been made by the aforementioned date, Lagoven may extend the deadline for making such decision. However, during the term of such extension, Lagoven shall have full power to terminate the association agreement at any time (Lagoven, 03/1993, Condition 10).

In subsequent association projects, the validity of the respective authorizations would hold, prior to their signature, for an indefinite duration. Moreover, the reckoning of the 'fixed duration' would not start at their date of signature but only at some subsequent event, for example, the first commercial shipment, and the term could be extended in the event of force majeure or other similar events. Nevertheless, thereafter all association agreements were to have a maximum duration of between 39 and 45 years.

But in the Cerro Negro AA, the last of the four Upgrading Association Projects, it was already anticipated that:

> [...] in the event of a change in Venezuelan law that allows this Agreement to have an indefinite term, the term of the Agreement will be automatically extended until the exhaustion of the Designated Area (Lagoven, 10/28/1997, Clause 16.1).

4.4.3. Special Cases and Cases of Public Interest

According to article 5 of the Nationalization Law, the AAs were to be limited to 'special cases, and whenever it so suits the public interest'. However, as regards the exploration and production of gaseous or liquid hydrocarbons, from extra-lights (condensates) to extra-heavy (Orinoco Oil Belt), PDVSA had all the required experience. Thus, for example, with regard to the Maraven-Conoco Project, the first to be introduced in the National Congress, in April 1993, Maraven stated that:

> PDVSA has extensive operational know-how in heavy crudes and in the production and transportation of extra-heavy crudes. In this activity, the association would use the techniques that the affiliate Maraven has been applying for more than 30 years.

In contrast:

> The conversion of crude from the Belt into upgraded crude is economically possible through the delayed coking technology in which Conoco has proven and solid experience (Maraven, 04/1993, Annex B, Relevant Circumstances, p. 13).

Hence, the upgrader represented a 'special case'; as for the Cristóbal Colón, it would be the liquefaction of gas. Remarkably, at no time did a proposal arise – whether coming from PDVSA, or from the government, or from the political parties – that the association be limited to the upgrading or liquefaction plants.

Then, in 1995, the Profit Sharing Association Projects were to be justified as "special cases" because they were to *accelerate* the exploration and exploitation of new areas in the face of a growing demand. Finally, in 1997, upon approval of the FOC of the third Upgrading Association Project – the Hamaca Project, promoted by Corpoven – in the Recitals the association was justified as follows:

> [...] with the incorporation of the participation by international associates, resources of the Venezuelan state are freed up and can be devoted to areas of higher priority [...], while *at the same time improving the balance of international payments* (Bicameral Commission of Energy and Mines, April 16, 1997; emphasis ours).

4.4.4. Association Agreements and Operative Agreements

Usually, in the oil industry, associations do not operate directly, but through an operating company, created specifically for this purpose and, of course, under their control. The operating company is generally remunerated on a 'cost plus' basis, covering production costs plus a conventional profit. The oil produced, at all times, remains the property of the association. However, all of these OAs were to be effectively controlled by the private partners due to the shareholding majorities of the latter. This meant that, consequently, the corresponding productive activities pursuant to the AAs were also privatized.

5. Authorized Association Agreements

5.1. Liquefied Natural Gas

As for the Cristóbal Colón project, there was only a *Preliminary Development Accord* signed by the partners in February 1991 in order to evaluate the feasibility of the Project. Thus, what was being requested was an authorization for a *hypothetical* Project, with the prospect of building a liquefaction complex with a capacity of 4.6 million tons per year (equivalent to 113 thousand b/d of oil). The investment within the national territory was estimated at 2.35 billion US$(1990). However, the authorization had an indeterminate duration and the partners never made it to the signing table for the AA. In fact, they would only do the absolute minimum while waiting for better times. Between 1991 and 2000, the partners were to spend only 50 million US$.

5.2. Upgrading of Extra Heavy Crudes

Regarding the Upgrading Projects, PDVSA's Board of Directors approved, in February 1993, a Framework Document of the Strategic Associations for the Exploitation and Upgrading of the Orinoco Belt Crudes and other Heavy and Extra Heavy Crudes (*Documento Marco de las Asociaciones Estratégicas para la Explotación y Mejoramiento de los Crudos de la Faja de Orinoco y otros Crudos Pesados y Extrapesados*). This document had been prepared "by a joint PDVSA and [MEM] team" and a copy was sent to the presidents of all the subsidiaries. They were also informed that this Framework Document had already served as a reference for the Maraven-Conoco Upgrading Project "which will soon be presented to Congress" (PDVSA, 02/18/1993).

Between 1993 and 1996, the National Executive presented four Upgrading Projects to the consideration of the National Congress. The four associations were to build the upgrading plants in Jose, located in the northern coast of Anzoátegui state, creating a new industrial complex. Now, although extra-heavy crude oil is a liquid under natural reservoir conditions, at the surface under standard conditions it solidifies, thereby *resembling bitumen proper*. For this reason, the extra-heavy crude would be diluted with naphtha, as it was being extracted, to obtain a liquid mixture, which would then be transported by pipeline to Jose. Finally, the upgrader would produce an upgraded crude, definitively a liquid, while the naphtha (which was originally imported) was to be recycled through smaller pipelines from Jose to the production field.

5.2.1. Maraven

The Maraven-Conoco Project – which would ultimately be called Petrozuata – was the first Upgrading Project presented to the National Congress, in April 1993, and was to be executed in the Zuata area of the Belt, assigned to Maraven. It would produce approximately 120 thousand b/d of 8° to 9° API extra heavy crude, which would be converted into 102 thousand b/d of upgraded crude (20° to 23° API). The proven reserves of the Zuata area were estimated at 10.6 billion barrels; the Project would only upgrade 1.5 billion barrels in 35 years, and would require an area of 200 km². The investment was estimated at 1.7 billion US$(1992) and the 'initial participation' in the capital of the Company was to be 49.9%, both for Maraven and Conoco, although the participation percentage could vary (Maraven, 04/1993). In the event, Conoco ended up with a 50.1% shareholding in the AA.

When the government of Carlos Andrés Pérez was toppled in June 1993, Maraven hurriedly introduced a second Upgrading Project Maraven-Total – which would ultimately be called Sincor –, which was essentially identical to the first one. It would involve 114 thousand b/d of extra-heavy crude to be converted into 100 thousand b/d of upgraded 31° API crude. Maraven would have an initial shareholding of 35%. Nothing was said, however, about the area required, nor about the amount of the investment (Maraven, 06/1993).

But in its *Report on the Study of PDVSA's* Strategic Associations on the MARAVEN-CONOCO and MARAVEN-TOTAL Projects [...] for the exploitation and upgrading of extra-heavy oils in the Orinoco Oil Belt (*Informe del Estudio de las Asociaciones Estratégicas de PDVSA sobre los Proyectos MARAVEN-CONOCO, y MARAVEN-TOTAL [...] para la explotación y mejoramiento de petróleos extrapesados de la Faja Petrolífera del Orinoco*), the Bicameral Commission referred, in both cases, to an area of 250 km². Moreover, it was noted that "the Zuata area has 10.6 billion barrels of proven reserves, and the projects would require only 3 billion barrels over a lifespan of 35 years. (Bicameral Commission of Energy and Mines, 12/08/1993, pp. 8-14).

None the less, in 1995 the MEM assigned an area of 299.5 km² to Petrozuata (GOV, 03/11/1995). In the AA, this area was divided into a base area of 230.5 km², and a reserved area of 69 km². The Project was to be developed in the base area, with a view to producing and upgrading 120 thousand b/d of extra-heavy crude which, in 35 years, would amount to 1,533 million b. However:

> The current Total Petroleum Requirement is 1533.0 million barrels, but this amount will change throughout the term of this Agreement as oil is produced

> and as changes are made to the annual production rate, production method, or upgrading plant capacity, as provided for by the revised Business and Investment Plan then in effect (Maraven, 1995, AA Definitions).

That is, after production of the extra-heavy oil started, the partners would have to decide whether or not to incorporate the reserved area. Since, according to Maraven, 200 km² were sufficient to support a production of 120 thousand b/d of extra-heavy oil, the allocated area of 299.5 km² would have been enough to support a production of approximately 180 thousand b/d.

On the other hand, the area assigned in 1997 by MEM to Sincor was 504 km² (GOV, 19/11/1997). In the AA this area was divided into a 'base area' of 324 km² and a 'reserved' area of 180 km², and it was specified that the "Total Project Oil":

> […] means two thousand and five hundreds and thirty […] barrels of Extra Heavy Oil expected to be produced […] or such other number of barrels of Extra Heavy Oil as the Participants may agree in accordance with this Agreement […] (Maraven, 1997, AA Definitions).

Sincor thus at once targeted a production of 198 thousand b/d, although in its documentation Maraven had only referred to a production of 114 thousand b/d. The potential of the allocated area could support a production of approximately 300 thousand b/d.

The authorizations for the two projects were published in the same Official Gazette, together with that of the Cristóbal Colón project, with essentially identical *Recitals*.

5.2.2. Corpoven

The Hamaca Project – as it was generally known, although the AA formally referred to Ameriven – contemplated the production and upgrading of 197 thousand b/d of extra-heavy crude (9° API gravity), which would be converted into 181 thousand b/d of upgraded crude (about 25° API). The total investment was estimated at 3.5 billion US$ (Corpoven, 11/1996).

A notable innovation was introduced in Condition 10 of its framework:

> […] for a period with a maximum duration of four years, the Association, in order to reach an optimum production level by the time the […] Upgrading Plant starts commercial operation, […] may undertake the production of extra heavy crude (the "Development Production") […]. The volume of such production shall be specified in the [AA]. The Development Production shall be sold to Corpoven who shall market it in the international markets, […] as a blend with other hydrocarbons suitable for transportation and handling, at market price […]. In addition to the Parties' provision of funds for the Project, the net cash flow resulting from the sale of the Development Production will be used to finance […] the Project.

The AA specified that this would be an average of 35 thousand barrels per day. Evidently, by generating a significant early revenue stream, the profitability of the project would be substantially enhanced. Moreover, the association would have full rights over the associated natural gas and other by-products obtained in the upgrading, including the right to export such by-products, which had not been the case with Petrozuata and Sincor.

As for the area, the FOC only refers to it in general terms, in Condition 1 (GOV, 20/05/1997). Later on, at Corpoven's request, the MEM assigned the project an area of 659 km² (GOV, 12/09/). In the AA, this was divided between a base area and a reserved area of 463 km² and 196 km², respectively. CVP's initial participation was 30% (Corpoven, 07/09/1997).

5.2.3. Lagoven

The Cerro Negro Project contemplated the production and upgrading of 120 thousand b/d of 8.0° to 8.5° API extra-heavy crude to 108 thousand b/d of upgraded 16° API gravity crude. Development production, prior to the start-up of operations at the Jose Upgrader Complex, could reach 60 thousand b/d. The total investment was estimated at 3.5 billion US$, with a national component of 1.94 billion US$. Lagoven and Mobil would have an initial participation of 41⅔% each, and Veba the remaining 16⅔%. The production corresponding to Lagoven and Mobil would be destined for the Chalmette Refinery, in which both were to participate with 50%. In total, this would amount to 90 thousand b/d, and the remaining 18 thousand b/d of upgraded crude would be destined for the Veba refining system, in Western Germany (Lagoven, 1996). In addition, it was contemplated "to develop the Project in an area of approximately 295 square kilometers" (Lagoven, 1996, p. 12).

In the approved FOC, Condition 1, it was only specified that the Project would:

> [...] exploit extra-heavy crude oil reserves located in the Orinoco Belt in the state of Anzoátegui, in a portion of the geographic area that has been assigned to LAGOVEN by the [MEM] (the "Determined Area") (GOV, 10/06/1997).

Now, in the Cerro Negro AA, in its Clause I: *Definitions*, it was specified that:

> "Designated Area" shall mean the geographic area in the Cerro Negro area of the Orinoco Belt [...], specified in Annex A and subject to the terms of the Reservation and Dedication Agreement having the same date as this Agreement.

The annex consisted of a single page with a sketch, identifying 295 km² within the Cerro Negro area, although in reality it only represented the base area, whereas the reserved area was to be determined by the Reserve and Dedication Agreement, which was not among its annexes. This Reserve and Dedication Agreement made reference to the:

> *Relinquishment of Portions of the Designated Area*: […] the participants of the association shall relinquish in favor of Lagoven any external and contiguous blocks of the Designated Area that the Parties unanimously determine, […] which are not necessary for the joint venture participants to economically achieve the agreed production goals of the project. The Parties may also by unanimous consent agree to expand the Designated Area (Lagoven, 28/10/1997, Reservation and Dedication Agreement, Clause 2.2; emphasis in original).

Consequently, the designated area was 2,814 km², i.e. Lagoven's entire area in the Belt, which was divided between 295 km² of base area and 2,519 km² of reserved area. In fact, the area was never published as such in the Official Gazette, and Lagoven assigned it directly to the Cerro Negro association: a fact without historical precedent.

Now, according to the AA, *Clause VIII*, once a production of 120 thousand b/d of extra-heavy crude oil was reached and construction of the upgrader was complete:

> […] one or more of the Parties may, without requiring a unanimous decision of the Parties […] provide additional funds for the Project for the purpose of increasing the Upgrader's capacity or increasing the production of Extra Heavy Oil.

Given the proven reserves of the area, which were approximately 12 billion barrels, it was perfectly possible to reach a production of one million barrels per day. And it is worth recalling that in this AA it was already foreseen that "in case of a change in the Venezuelan law that allows this Agreement to have an indefinite term, the term of the Agreement will be automatically extended until the exhaustion of the Designated Area".

Moreover, the National Congress, in authorizing the AA, considered that:

> […] the Bicameral Commission of Energy and Mines of the Congress of the Republic has carried out a detailed analysis of the contents of the [AA] submitted by the [MEM], having verified that said [AA] is in accordance with the [FOC] approved by the Congress of the Republic on April 24, 1997 […].

Therefore, it was agreed:

> Article 1. To authorize the execution of the [AA] *and its annexes* […] (GOV, 10/15/1997; emphasis ours).

Thus, even the execution of the annexes, which were never put before the Bicameral Commission, was authorized.

5.3. Orimulsion

In 1996, the National Executive presented to the National Congress Bitor's first Association Project with Conoco and Statoil (GOV, 17/07/1996). The objective of this association was to convert around 80 thousand b/d of extra-heavy crude into Orimulsion, for a long-term supply agreement with Florida Power & Light, a power generator located in Florida (USA). Already in 1993 PDVSA had renamed the Orinoco *Oil* Belt as the Orinoco *Bituminous* Belt and now, among the Recitals of the authorization of the FOC by the National Congress, it was stated that "in the Orinoco Belt there are substantial proven reserves of natural bitumen". Then, in the FOC, in its Condition 11, it was stated that:

> By virtue of the fact that [...] Orimulsion is a non-petroleum hydrocarbon, the production levels of natural bitumen [...] shall not be considered subject to the international commitments of the Republic of Venezuela derived from its participation in international organizations.

This Condition 11 did *not* refer to OPEC quotas, which are based, as befits a conservationist policy, strictly on the production levels of the natural resource, and not of derivatives. In fact, the idea was that Orimulsion should be recognized by the *importing* countries as a 'liquid coal', but coal nonetheless, and therefore not subject to the policies aimed at minimizing the import of crude oil and products from OPEC member countries. Indeed, this was the position finally adopted by the main consuming countries, so that Orimulsion could be imported freely, just like coal in general.

Bitor, for its part, guaranteed the supply of Orimulsion in the required quantity, so that the very significant investment in abatement equipment to reduce the pollution of the atmosphere and bring it to tolerable levels would be economically justified. This commitment, in turn, was backed outside of Venezuela by PDVSA. If, for any reason, Bitor were unable to deliver the required volumes, these were to be replaced by residual fuel at *an equivalent price*, from any source.

Moreover, in August 1996, the Ministry signed a new royalty agreement with Bitor, pursuant to which the price of Orimulsion was decoupled from the price of heavy crude oil and was set at a minimum of US$ 0.682/b. According to the 1943 Hydrocarbons Law, article 50, this was the absolute minimum value established for the royalty liquidation price. With the rate stipulated at one-sixth, Bitor was to pay, as a minimum, the amount of US$ 0.11/b. However, at the same time a *ceiling* – a true innovation – was also established, at a liquidation price of 7.042 US$/b, which meant that Bitor would pay at a maximum 1.17 US$/b (MEM, 08/1996). In 1997, the first year the royalty agreement was in effect, Bitor's royalty liquidation price for a barrel of extra-heavy crude averaged 0.71 US$, and the royalty per barrel averaged US$ 0.12 US$; whereas a barrel of extra-heavy crude blended with lighter crudes, its average price was 13.01 US$/b, which was equivalent to a royalty of 2.17 US$/b. Moreover, the Orimulsion AA was to be subject to the non-oil income tax rate of 34% (Boué, 2012, p. 200).

However, this association did not come to fruition as Florida Power & Light opted for natural gas, more expensive than Orimulsion, but much less polluting.

6. Tax Regime for Associations

6.1. Income Tax and Fiscal Export Value

PDVSA argued that both the Cristóbal Colón Project and the Upgrading Projects were of marginal profitability and that, consequently, the applicable income tax rate should be the non-oil rate, and neither of them should be subject to the VFE. In the words of Minister Celestino Armas:

> It is important and opportune to highlight, due to its decisive weight, the particular attention given to tax matters. The fate of the project will depend on the decisions taken in this area. To apply to it the prevailing taxation regime for hydrocarbons would be tantamount to foregoing with its implementation (Armas, 14/06/1991, p. 7).

He then referred to the need to reform the Income Tax Law accordingly. However, Celestino Armas did not say anything about the royalty rate. Indeed, in all matters related to income tax and VFE, a legal reform was required to make any change, whereas with respect to the royalty rate, according to the 1943 Hydrocarbons Law, the government had the legal faculty to reduce it.

In 1991, with the reform of the Income Tax Law, the associations whose business involved *integrated* activities (i.e. activities spanning the primary sector of extraction and the secondary sector of industrial processing), like those engaged in the production of natural gas and its liquefaction, or in the production of extra-heavy crude oil and its upgrading, for example, were to benefit from the reduction of the tax rate to the non-oil level (GOV, 08/13/1991). However, until June 1993, they would all continue to be subject to the Fiscal Export Value. In that month, the National Congress approved the immediate elimination of the VFE for all associations and its gradual reduction to zero for PDVSA and its affiliates by 1996 (GOV, 30/06/1993). From 1996 on, PDVSA was to pay, for the first time in its history, dividends.

6.2. Royalty

In 1943, natural gas was not a source of international ground rent since, perforce, its consumption was domestic. For this reason, the Hydrocarbons Law allowed the government to reduce the royalty rate to zero, contractually and for a period of up to fifteen years (art. 41). The Cristóbal Colón Project aspired to a similar arrangement, as stated in the draft AA sent to the National Congress. The partners were to "negotiate and secure a satisfactory accord with the [MEM] regarding the applicable exploitation tax" (Lagoven, 03/1993, p. 47).

On the other hand, since crude oil was a source of an international ground rent, the 1943 Hydrocarbons Law allowed the National Executive to reduce the royalty rate only "in those cases in which it is demonstrated to its satisfaction that the cost of production [...] prevents the commercial exploitation of said substance", but it also allowed the Executive to "raise again the rebated tax, until re-establishing it at its original level when, in its judgment, the causes that motivated the reduction have ceased" (art. 41). In turn, according to Maraven's Project regarding Petrozuata, this article would be given "special attention in the case of projects for the development and production of extra-heavy crude oil from the Belt" (Maraven, 04/1993, p. 29).

In May 1998, PDVSA Petróleo y Gas – a subsidiary which, in December 1997, arose from the merger of the operating subsidiaries of PDVSA – signed a Royalty Agreement with MEM corresponding to the Upgrading Projects. The royalty rate applicable to the 'Development Production' – that is to say, oil produced before the start-up of the upgrader – would still be one-sixth. But upon commencement of upgrading, this rate was to be reduced to one percent for a period of nine years as a maximum, provided that the gross income of the association was less than three times the total investment. Then, in principle, the one-sixth rate was to be reinstated, described as the "maximum currently allowed by the Hydrocarbons Law" (MEM, 1998). For their part, all that the private associates had to do was to adhere to the Agreement in question by means of a letter, and they did so.

6.3. Conclusions

As part of the "Oil Opening" of the 1990, PDVSA worked out a draft project to implement a non-proprietorial tax regime throughout the oil sector. The statutory royalty rate was to be reduced to a maximum of 5%, and the income tax rate – "upstream and downstream" –, would be 34%, just like in the non-oil sector. Finally, in "upstream" activities, an excess profit tax was to be created (PDVSA, 15/03/1997).

7. Privatization of the Equity Participation of Subsidiaries

The shares of PDVSA's subsidiaries could not be privatized as long as they were linked to a guarantee obliging them to indemnify, if necessary, the other associates. No private partner would take on such a risk; hence, if the affiliates were to be privatized, they would first have to be somehow released from this obligation. In the Cerro Negro AA, the way to do this was already spelled out. Lagoven:

> [...] shall have no obligation to compensate a Foreign Party for damages suffered [...] as a result of any Discriminatory Measure taking place subsequent to the Venezuelan state reducing its interest [...] to (i) less than 12.5% in the Project or (ii) less than 49.9% of Lagoven (Lagoven, 28/10/1997, Clause 15.2).

However, this reduction of the Venezuelan state's participation in the association, or in Lagoven, could not be carried out without Mobil's written consent (Lagoven, 28/10/1997, Clause 13.1); and, of course, it would not consent without a guarantee that these 'Discriminatory Measures' would somehow be compensated.

8. National Congress and Oil Opening

The promoters of the Oil Opening had done everything possible to avoid any political discussion. But in the midst of the agitated political environment, with the fall of Carlos Andrés Pérez and the general elections in sight, such discussion nevertheless arose – surprisingly – revolving around the Cristóbal Colón Project.

Deputy Alí Rodríguez Araque (*Causa Radical*) emphasized that no private investor would give guarantees similar to those being assumed by PDVSA vis-à-vis foreign investors. In fact, Lagoven's only role was to guarantee free access of foreign investors to the natural resource. Given the low profitability of the Project, Rodríguez Araque wondered what explained the interest of Lagoven and its associates in it, not to mention the extreme haste of Congress to approve this Project as soon as possible:

> [...] it is about establishing a legal precedent, anointed by this Congress, according to which it is no longer the state that exercises control over the natural resource, as we moved from the old system of concessions to a worse one [...] (Rodríguez Araque, 08/03/1993).

Deputy Alvaro Silva Calderón (*Movimiento Electoral del Pueblo*) was the only member of the Bicameral Commission that emitted a dissenting vote against the Report as presented, and he was also the only one in the National Congress to refer to the BIT with Holland and to place it in the corresponding context. Although the FOC stated that the AA would not engage the responsibility of the Republic, Silva Calderón warned that "a series of treaties for the protection of foreign investments is being proposed to Venezuela, so that companies such as these would receive preferential treatment". For Silva Calderón, in short, it would definitively "be admitted as a commitment of the Republic" (Silva Calderón, 1993). The political atmosphere at the time was charged with rumors of coups d'état, but the Congressman insisted that:

> It would be very difficult even for Parliament itself to make modifications to these legal texts, including in the eventuality of a de facto government, because when governments of this kind come into being, they begin by declaring that they recognize extant international commitments (Silva Calderón, 1993).

Silva Calderón had no doubt that Lagoven and its associates were committing themselves to very little, while the Republic was dismantling its entire legal, economic and political structure, which dated back to the concessionary regime. A simple concession would have been more advantageous for the country.

Nevertheless, all the participants in the parliamentary debate, including those from the opposition ranks, felt compelled to explicitly recognize the achievements of PDVSA and its subsidiaries, including the Internationalization policy, and to express their agreement, *in principle*, with the Oil Opening.

The National Congress ended up approving the FOC originally proposed by PDVSA, and it was approved not only with the votes of the two main parties, AD and Copei. The project also had the votes of all the Senators and Deputies of the state of Sucre, where the Cristobal Colón was to be undertaken. The illusion promoted by PDVSA/Lagoven of a very large investment that would result in an extraordinary economic boost for this state, one of the poorest in the country, was most successful.

9. International Energy Agency, OPEC Quotas and Orimulsion

In 1998, Venezuelan Minister Erwin Arrieta incited a confrontation within OPEC because the Organization accused Venezuela of exceeding its quota by one million barrels per day. Arrieta denied the fact, despite ample evidence to the contrary, and stated that the system of quotas determined on the basis of official information provided by member countries could not continue as it was, since all of them lacked credibility. As a solution, he demanded that data should be used from supposedly independent 'secondary sources', which published their own estimates. Arrieta prevailed and OPEC selected six 'secondary sources'. Since then, the average of the estimates from these secondary sources are the ones that serve as the basis for determining OPEC quotas, whereas the official figures published by the respective governments are ignored. Among these six sources was the IEA, which in practice was to serve as a reference for the entire group. This meant that, definitively, the approximately 75 thousand b/d of Orinoco Oil Belt extra-heavy oil – processed by Bitor in Orimulsion – would no longer be part of the OPEC quota, since the IEA did not account for them as part of Venezuela's output of crude oil.

10. Hugo Chávez: From the IV to the V Republic

In the December 1993 elections, the votes of AD and Copei – and of their presidential candidates – came to only 46% of the total, a percentage which, in itself, was an indicator of an ongoing political crisis. The Presidency was won, for the second time, by Rafael Caldera – already separated from, and running against, Copei – with the support of the recently founded *Convergencia Nacional* party. He obtained 30% of the votes, although *Convergencia Nacional* barely came to 17%, a result that placed it third, behind AD and Copei. The foundations of the prevailing political regime were giving way. Real wages continued to deteriorate and tax levels in the non-oil sector continued to increase. But PDVSA and its subsidiaries, with the wind in their sails, were prospering and enjoying extraordinary power.

However, 1998 witnessed the worst price collapse in the world petroleum market since 1929, attributable mainly to PDVSA's policy of ignoring OPEC quotas. The country had already exceeded its quota by one million barrels per day and PDVSA continued to proclaim to the four winds that it was not going to cut production by a single barrel, while also insisting on the country's exit from OPEC. Meanwhile, twenty years after GDP per capita began to shrink, the Venezuelan economy collapsed again. In January 1998, an election year, Irene Sáez – political scientist and former Miss Universe –, headed the polls of presidential voting intentions. Hugo Chávez was also a candidate – having been pardoned by Rafael Caldera –, although he was not yet a major contender, and did not even figure in the first polls of that year. But by December, with oil prices on the floor and the public treasury in ruins, Hugo Chávez won the presidential elections with 56% of the votes.

For their part, PDVSA's top executives, with Luis Giusti as president, had prepared themselves to assume positions of great relevance in the event of a new government presided by Irene Sáez, convinced that the definitive culmination of the Oil Opening was within reach. However, while totally confident of their success at the beginning of the year, by the end of it all they could do was witness the collapse of the Fourth Republic (Arrioja, 1998).

Chávez, in his electoral program, stated that the political regime, which emerged from the *Punto Fijo* Pact in 1958, had already reached its limits by the end of the 1970's, which led to "the collapse of the state, the impoverishment of the population and political discredit". According to Chávez, this regime was beyond repair. He therefore announced a "Constituent Process", "a democratic revolution because it unleashes the profound transformation of Venezuelan society, based on the popular will, by exercising its sovereignty" (Chávez, 04/1998, p. 3).

As to oil policy, he limited himself to the criticisms characteristic of any opposition candidate. He only stood out for advocating, in the midst of the price collapse, for the strengthening of "relations with OPEC in order to reinstate control of supply" (Chávez, 04/1998, p. 27). However, these quotas would continue to be based on 'Secondary Sources'.

10.1. Investment Promotion and Protection Law

In the context of a serious economic and financial crisis that required urgent actions, the National Congress – the government was a minority one – approved an enabling law in April 1999 (GOV, 04/26/1999). Based upon it, the government enacted, by decree, the Investment Promotion and Protection Law (GOV, 10/22/1999). Its purpose was:

> […] to provide investments and investors, both domestic and foreign, with a stable and predictable legal framework, in which they can develop in an environment of security, regulating the actions of the state towards such investments and investors […] (Art. 1).

What was to be understood by 'investments' was specified in a long list, which included:

> […] rights obtained under public law, including concessions for exploration, extraction or exploitation of natural resources […] (Art. 4).

The Law, in turn, ratified the validity of existing investment promotion and protection treaties, and authorized the government to:

> […] enter into legal stabilization contracts, for the purpose of ensuring the stability of certain economic conditions for the investment throughout the term of such contracts. These […] may guarantee to the investment one or more of the following rights:
>
> […] Stability of the national tax arrangements in force at the time the contract is entered into (Art. 17).

These contracts would require the authorization of the National Congress and, for the time being, would only have a duration of ten years.

10.2. Bolivarian Constitution

Chávez also immediately called for a National Constituent Assembly, which was approved by referendum in April 1999. In the corresponding elections of July 1999, his followers obtained an absolute majority. The National Constituent Assembly sat for six months, and the new Constitution was finally approved by another referendum in December 1999. With the new Constitution, following President Chávez's personal proposal, the name of the country was changed from Republic of Venezuela to *Bolivarian* Republic of Venezuela (BRV). The new Constitution referred to petroleum in two of its articles:

> Article 302: The state reserves to itself, through the pertinent organic law, and for reasons of national expediency, the petroleum industry […].

The Nationalization Law was thus validated. Consequently, the reinterpretation of the Nationalization Law by the National Congress throughout the Oil Opening process was also maintained. Moreover:

> Article 303: On grounds of sovereignty […], the state shall retain all of the shares of Petroleos de Venezuela, S.A. […], except the shares of subsidiaries, strategic joint ventures, business enterprises and any other venture either established or to be established in the future as a consequence of the development of businesses of Petróleos de Venezuela, S.A.

That is to say, 'Petróleos de Venezuela, S.A.', the *holding* company, which did not produce a single barrel of petroleum, nor a single cubic foot of natural gas, as an entity regulating petroleum as a productive activity had obviously to be exclusively owned by the state. But the affiliates that carried out the productive activities could be privatized.

Finally, in July 2000, new general elections were held. Chávez obtained 60% of the votes, and now had a majority in the National Assembly.

11. Oil Policy (1999-2004)

Hugo Chávez, upon assuming the Presidency, left no doubt that he had been elected President of all of Venezuela, both of the non-petroleum sector and the petroleum sector. PDVSA was a corporation and he considered himself as its sole shareholder, in representation of the state. Luis Giusti and his immediate collaborators had to leave PDVSA. Chávez appointed Alí Rodríguez Araque as Minister of Energy and Mines, who immediately ordered PDVSA to put an end to the Cristóbal Colón Project. He also initiated a process to recover fiscal control over PDVSA, as to export prices and royalty liquidation prices.

On the other hand, Chávez, on the recommendation of Alí Rodríguez Araque, also appointed as president of PDVSA Roberto Mandini, a senior industry executive who had always opposed Luis Giusti. But Chávez also appointed Héctor Ciavaldini (a chemical engineer, former PDVSA employee, and main advisor in petroleum policy to Chávez when the latter was a candidate) to PDVSA's Board of Directors. Ciavaldini became, *de facto*, PDVSA's Vice President. Confronting Roberto Mandini, he forced the latter's departure in August 1999, and Chávez appointed him President of PDVSA. Ciavaldini, a staunch defender of PDVSA as the entity in charge of petroleum policy, convinced Chávez to exclude the Minister from any important meeting in Miraflores. Furthermore, Ciavaldini disqualified all initiatives coming from the Ministry regarding fiscal control over PDVSA, as being redolent of corruption. Ciavaldini was also an advocate of Orimulsion, whereas Alí Rodríguez Araque was firmly opposed to its development.

At Chávez's initiative, in September 2000, the *Second Summit of Heads of State and Government of OPEC Member Countries* was held in Caracas. Rodriguez Araque took advantage of the opportunity to become a candidate to the OPEC General Secretariat, which he assumed in January 2001 and, for the time being, he kept distance from national oil policy, and Chavez now appointed Alvaro Silva Calderón as Minister of Energy and Mines.

In October 2000, Chávez dismissed Ciavaldini, who had proven to be incapable of managing a company of PDVSA's complexity. He now appointed as president General Guaicaipuro Lameda, who had excelled as an administrator of public finances of the government, but had no experience related to petroleum.

11.1. New Hydrocarbons Legislation

11.1.1. Arbitration

The above mentioned Enabling Law authorized the President to decree a new Organic Law of Gaseous Hydrocarbons, with which non-associated natural gas was completely opened to private investment (GOV, 11/23/1999). Moreover, for the first time in the history of Venezuelan hydrocarbons legislation, this new law referred to arbitration as part of an "amicable" settlement:

> Any doubts and controversies of whatever nature that may ensue because of this licence and which cannot be amicably settled, *arbitration included*, shall be decided upon by the competent courts of Venezuela, and in accordance with its laws, and for no reason nor for any cause shall they give rise to foreign claims (Art. 4; emphasis ours).

In 2001, with Alvaro Silva Calderón as Minister of Energy and Mines, another enabling law authorized the President of the Republic to issue a new Organic Hydrocarbons Law (GOV, 11/13/2001). This Law continued referring to arbitration as an amicable settlement mechanism.

11.1.2. Royalty and Income Tax

The new Organic Hydrocarbons Law came into force on January 1, 2002. On the one hand, it validated (while also making subordinate) the Organic Law of Gaseous Hydrocarbons, and on the other hand, it repealed the 1975 Nationalization Law. The Organic Hydrocarbons Law set the usual royalty rate at 30%, although the National Executive, at its sole discretion, could lower this rate to 20% and, in turn, raise it again (art. 44).

The royalty rate increase came just in time. The income tax legislation had permitted the oil companies to account for the consequences of the devaluations of the bolivar as losses, manipulating their accounts receivable between dollars and bolivars. PDVSA had also brought down its effective income tax rate and, with the 2002 devaluations, it was reduced to zero. In contrast, the royalty is originally a percentage of gross income in dollars.

At the same time, the Income Tax Law was reformed. On the one hand, the oil tax rate was reduced from 67.7% to 50% but, on the other hand, it put an end to the aforementioned accounting manipulations.

11.2. Bitor and the Organic Law of Hydrocarbons

Although published in the Official Gazette of November 13, 2001, the Organic Hydrocarbons Law was only to become effective on January 1, 2002. However, in great haste, due to a commitment made by President Chávez in his first visit to China in 1999, the National Congress authorized, on December 17, 2001, a Bitor Association Project with the *China National Oil and Gas Exploration and Development Corporation and Petrochina Fuel Oil*

Company Limited – ultimately called Sinovensa –, which was to be governed by the old legislation still in force. The FOC was identical to the one already approved in 1996 in connection with Bitor's Association Project with Conoco and Statoil, and the AA was signed on December 27, 2001, with Bitor's initial participation at 30%. All Orimulsion production was to be exported to the People's Republic of China (GOV, 12/17/2001).

11.3. PDVSA and the Organic Law of Hydrocarbons

The Organic Hydrocarbons Law was a summary of what had been the most advanced legislation during the concessionary regime. Moreover, in all primary activities – defined as exploration, exploitation, gathering, transportation and initial storage of hydrocarbons –, the state would always have a majority shareholding (art. 9), but 'downstream' activities were fully opened to private investment. AAs were to have a maximum duration of 25 years, and the corresponding areas would be assigned to them by Presidential Decree.

With the new Organic Hydrocarbons Law, the MEM was formally going to recover, vis-à-vis PDVSA and its subsidiaries, the functions it had during the concessionary regime vis-à-vis foreign concessionaires. However, throughout the concessionary regime, the Venezuelan public opinion had identified the national oil policy with the Ministry of Mines and Hydrocarbons, which translated into widespread popular support and, therefore, also the support of the National Congress. In contrast, since the nationalization, public opinion identified the petroleum policy with PDVSA and its affiliates, and no longer with the MEM.

The President of PDVSA, Guaicaipuro Lameda, publicly opposed the new Organic Hydrocarbons Law. He had been a member of the commission in charge of drafting the corresponding Bill and, once the drafting of the Bill was complete, Lameda summarized the reasons why he was opposed to it. He recalled that, as a military, he had sworn to "honor courage, honesty and truth, which I will defend even at the risk of my safety and well-being", and affirmed that he now shared "similar concepts with those who make up the petroleum industry because, for the latter, they constitute a corporate endowment that has driven it since its birth" (Lameda, 11/08/2001).

But in the commission, Lameda had come face to face with "experts and university professors", who had been part of the 1974/5 Reversion Commission, and had in vain opposed the governmental Nationalization Law Project. But now, with a majority in the commission in charge of the Bill:

> […] we went through all of the events in the past that annoy or bother some members of the Commission. […] I heard opinions that we must do something to prevent PDVSA from perverting its fiscal contribution; to prevent it from imposing its will on the government; to recover its control because it is said to be extremely powerful and uncontrollable, and that it has gotten out of the

hands of the Ministry it was attached to. In my opinion these perceptions do not reflect the reality of the corporation that I am honored to lead today [...].

General Lameda's vision was different:

> I am convinced that [...] the energy demand of today's world offers us an extraordinary opportunity for the country to have an energy corporation which would be a global reference and which, with its excellence, would add the maximum possible value to the Venezuelan nation.

In particular, he disagreed with the royalty. He insisted that it should be flexible, according to the characteristics of each field. He also disagreed with the majority shareholding in primary activities, and insisted that the new law should include an explicit guarantee that it would not apply to existing operative and association agreements which had been based on the Nationalization Law. Finally, he referred to:

> Governance and the distinction between the role of state-owned companies and the centralized organization of the [MEM].
>
> This issue was fundamentally addressed from the historical perspective of an unsatisfactory relationship between PDVSA and the MEM.
>
> [...] I perceive that when talking about the governance of productive and commercial activity we are comparing two strategies: division in order to govern versus strengthening [PDVSA] to add value.

He had no doubt that, as far as PDVSA was concerned, it was convenient to "strengthen it in search of the greatest possible addition of value for the nation".

11.4. Coups d'état

In February 2002, Chávez dismissed Lameda and appointed as his successor Gastón Parra Luzardo, at the time Vice-President of the Venezuelan Central Bank (and one of the professors to whom Lameda had alluded in his speech). A part of PDVSA's executive and senior staff did not want to accept his appointment, and some even manifested their sentiments publicly. Chávez responded, also publicly, by ordering their dismissal.

In addition, the country was in a state of general agitation. On a political level, the opposition concentrated on denouncing Hugo Chávez's personal rapprochement with Fidel Castro, and the country's rapprochement with Cuba. On the other hand, the Hydrocarbons Decree-Law was only one of 49 decree-laws published on the same date – November 13, 2001 –, which the Venezuelan business establishment and the Federation of Chambers and Associations of Commerce and Production (Fedecámaras) considered unacceptable because they minimized the role of the private sector. The

Venezuelan Workers Confederation, for its part, remained in the hands of the traditional political forces and acted in coordination with Fedecámaras. Finally, on April 10, 2002, both sectors united in calling for an indefinite general strike, not only demanding the annulment of all decree-laws, but also the resignation of President Chávez and the celebration of new general elections.

On April 11 a massive demonstration – led, among others, by General Guaicaipuro Lameda –– approached Miraflores (the presidential palace) which, in turn, was surrounded by a massive demonstration in support of Chávez. That day 18 people were victims of snipers in circumstances that were never clarified. The following day, in the early morning of April 12, a military coup d'état took place and President Chávez was arrested. On the same day, the President of Fedecámaras, Pedro Carmona, was sworn in at Miraflores as President of the Republic. His first official acts included the dissolution of the National Assembly, the dismissal of all governors and ambassadors, and the elimination of these 49 decree-laws, including the Organic Hydrocarbons Law. Moreover, he reinstated General Guaicaipuro Lameda as President of PDVSA.

But the massive mobilizations in support of President Chávez continued, and a military counter-coup returned him to power. On April 14, Chávez came back to Miraflores, by helicopter, where his entire cabinet was waiting for him amidst a huge crowd of his followers.

Upon his return, President Chávez took a prudent position regarding PDVSA. He asked Alí Rodríguez Araque to return to Venezuela to replace Gastón Parra as President of PDVSA. Minister Alvaro Silva Calderón was now going to take office as Secretary General of OPEC. The new Minister of Energy and Mines would be Rafael Ramírez Carreño, an engineer who, until then, had been the president of the *Ente Nacional de Gas*, created pursuant to the Organic Law of Gaseous Hydrocarbons.

Alí Rodríguez's main task was somehow to control PDVSA. However, his top management began to prepare, immediately and conspiratorially, a new coup, which this time was to be a petroleum coup, in order to force the resignation of President Chávez. On December 2, 2002, it initiated a petroleum strike – in fact, a lockout –, and thousands of PDVSA employees abandoned their jobs. However, the effectiveness of the lockout was primarily due to the blockade carried out by the PDV Marina fleet to prevent the departure of vessels from the ports of export. In fact, the workers in the production fields did not participate in the conflict and only stopped production when there was no more storage capacity left. By the end of December, the government succeeded, with spectacular military actions and with the personal participation of President Chávez, in clearing the ports. The strike weakened progressively and its promoters put an end to it on February 3, 2003.

This time, however, the participants in the strike were surprised to find that they were not allowed to return to their jobs. In short, of the 42,000 workers on the payroll of PDVSA and its subsidiaries, more than 40% were dismissed.

The events of 2002 and 2003 brought about a military presence in all of PDVSA's important installations and also in the internal structure related to the company's security. They were to have a say in the hiring of new personnel, filling vacancies and vetoing all those whom they considered enemies of the Fifth Republic.

Production in 2001 was 3.34 million b/d, a level that has never been reached again, despite new production of 620 thousand b/d of extra-heavy crude for upgrading purposes. In short, the forced shutdown of production also caused permanent damage in the order of some hundreds of thousands of barrels per day in the older traditional production areas, particularly in Zulia.

However, the opposition was not giving up yet. The Bolivarian Constitution provided for the possibility of revoking all those occupying popular election posts, once half of the respective mandate had passed, by means of a referendum, which could be activated if the requirement of gathering the corresponding signatures of 20% of the citizens registered in the electoral roll was fulfilled. Finally, on August 15, 2004, the Recall Referendum took place and President Chávez was confirmed with 59% of the votes.

11.5. International Energy Agency, OPEC Quotas and Upgraded Crudes

In July 2002, the IEA announced that it would also not consider upgraded extra-heavy crude as part of Venezuela's petroleum production. This, according to the IEA, reflected "the intent of the association agreements" (IEA, 12/07/2002, p. 22). Thus, the corresponding extra-heavy crude was not going to be part of the OPEC quotas either. In the short term, this represented another 300 thousand b/d, of extra-heavy crude, but within the next three years this output was slated to reach 620 thousand b/d.

In fact, this development had already been foreseen in the Upgrading AAs. Although the Petrozuata AA – signed in 1995 – did not refer to production quotas, the last one of the four – the Cerro Negro AA signed at the end of 1997 – specified that quotas would only apply to "the production of hydrocarbons of the types subject to the reduction" (Lagoven, 10/28/1997, Clause XIV). Consequently, Venezuela was to move away from OPEC, step by step, as it moved forward in the Orinoco Oil Belt, producing Orimulsion and upgraded crudes.

The situation was now as follows. The extra-heavy crude, processed into Orimulsion or upgraded crudes, would not be part of what the IEA considered as Venezuelan petroleum production, a position shared by all the Secondary Sources. Consequently, such crudes would also not form part of the Venezuelan quota agreed by OPEC. On the other hand, extra-heavy crude forming part of blends with lighter crudes, produced by PDVSA, would be included in the quota.

However, the government of President Chávez would continue to report Venezuelan petroleum production to OPEC, including extra-heavy crudes, and did so to underline that, in effect, the petroleum strike had been successfully overcome; whereas OPEC was going to report only the production subject to the Venezuelan quota determined by the Secondary Sources, and this was much lower, as it excluded the extra-heavy crudes processed into Orimulsion or Upgraded Crudes. The opposition, for its part, was going to insist on the latter, and accuse the government of lying about the true production levels.

12. Renegotiation of Operative and Association Agreements

Only after overcoming the Recall Referendum, more than five years after he took office, did the Chávez government manage to stabilize itself. He appointed the incumbent Minister of Energy and Mines, Rafael Ramírez, also as President of PDVSA.

Ramirez was determined to put right the operative and association agreements, which up until then PDVSA had refused to do. In effect, these agreements continued to be governed by the 1975 Nationalization Law, although PDVSA, being of the exclusive ownership of the state, was immediately subject to the new Organic Hydrocarbons Law.

For this specific purpose, in January 2005, Chávez appointed Bernard Mommer as Vice-Minister of Hydrocarbons. Mommer, who had been serving as the director of PDVSA's office in Great Britain, had there enlisted the help of an US-American law firm – *Curtis, Mallet-Prevost, Colt & Mosle LLP* – for certain matters related to Orimulsion. This law firm met a *sine qua non* condition to assist the Ministry in connection with these matters: It had never previously worked with either PDVSA or its affiliates. Therefore, this law firm had no involvement in the existing operative and association agreements.

12.1. Operative Agreements

The MEM considered that the OAs were illegal, since they were in fact hydrocarbon producing companies and, therefore, would have required the authorization of the National Congress. That said, they had been operating for twelve years without any such authorization, and five of those twelve years corresponded to the period in office of Hugo Chávez. Therefore, annulling them outright was not an option, but the government took the position that, being as their operations were unlawful, they would have to migrate to a new legal regime.

At the request of the MEM, the SENIAT – the agency responsible for administering the income tax – reviewed the income tax returns of the OAs. All had identified themselves as simple service companies and, therefore, subject to the non-oil rate of 34%. The SENIAT was now going to apply the oil rate to them as oil producers. According to the Income Tax Law, the accounts were kept open for review for a span four years, so the 67.7% rate was to be applied for the year 2001, and the 50% rate for 2002 and subsequent years.

While most of the OAs had correctly paid their income taxes and therefore only had to accept the adjustment in the applicable rate, several companies had never paid even a single cent in income tax, despite having significant production. On the basis of several reforms to the Income Tax Law, they accounted for their investments in Venezuela as loans from their parent companies, which meant that with each subsequent devaluation of the bolivar, they recorded losses large enough to never generate any taxable profits. The SENIAT intervened and reviewed their income tax filings in order to reach reasonable understandings, even though these could only cover the years from 2001 onwards (Ministry of Energy and Petroleum, 09/2005).

The solution offered to the OAs in order to bring them into line was their migration to joint ventures, in accordance with the framework of the Organic Hydrocarbons Law. In the negotiations, the MEM benefited from the close cooperation of the CVP, under the presidency of Eulogio Del Pino. By converting these OAs into joint ventures, the state was to acquire a 60% shareholding majority, and private investors were to be compensated according to the corresponding amount of their investment. The applicable royalty rate was to be 33.33%. A 3.33% was to be allocated to the respective municipalities where the production activities were located, because the municipalities had benefited from some contributions from the OAs, and the government did not want to antagonize them.

All OAs were terminated on March 31, 2006. Of the 32 operative agreements, 30 agreed to migrate or permanently relinquish their already obsolete agreements. In total, the government paid some five billion dollars in compensation, which represented less than half of the estimated net present value, as only the investment itself was recognized as being compensable. According to their original terms, the OAs were set to expire between 2013 and 2017. However, the term of the new joint ventures would be a further twenty years after the date of migration, that is, until 2027. The total area of 43 km² conceded was reduced to 27 km². The marginal take of the state, in its capacities as owner of the natural resource, as shareholder and as sovereign, came to 86.67% (royalty of one-third, income tax of 50%, and shareholding participation of 60%).

Only two OAs refused to migrate, Jusepín (Total) with a production of about 30 thousand b/d, and Dación (Ente Nazionale de Idrocarburi) with about 65 thousand b/d. In both cases amicable agreements were finally reached on the corresponding compensation, which came close to one billion dollars.

12.2. Association Agreements

The government did not question the legality of the AAs, because they had been authorized by the National Congress. However, it considered that these agreements were plagued by legal flaws. Rectifying this situation was then a matter of eliminating or accommodating these irregularities, on the one hand, and of *buying* an additional shareholding in order to achieve a majority in favor of the state as provided by law, on the other hand.

There were sixteen AAs in all, although here we will concentrate on the four Upgrading AAs, which were of major importance. Straight off, the MEM ordered these four associations to abide by their original production limits, in accordance with the Reports of the Energy and Mining Commissions of the National Congress. The associations were also informed that they had to abide by the base area defined in the AAs. However, as to production volumes, Sincor was an exception, producing and upgrading more than 200 thousand b/d of extra-heavy crude, when it had only been authorized to produce 114 thousand b/d. In view of this situation, the MEM decided to subject the excess production to a 30% royalty. With migration, Sincor was to be assigned an area of 399 km^2, although the Relevant Circumstances only referred to 250 m^2, but it had to pay a bonus of 130 million US$ for the additional 149 km^2.

In the case of Cerro Negro, a similar procedure was followed, with the recognition of an area somewhat larger than the 295 km^2 of the initial area, but they had to pay a bonus of 50 million US$. In the case of Ameriven, the area of 647 km^2 was simply reduced to 463 km^2.

The new joint ventures were to have the right to increase the production of extra-heavy crude to the extent that they also increased their respective upgrading capacities, which meant that they would all have an incentive to improve their recovery factors. In addition, all of them were authorized to use associated gas for the same purpose. Finally, they were also allowed to produce the minimum level of extra-heavy crude oil which would be technically required during the maintenance periods of the upgraders – the logic of the production process demanded it –, and to export such output as part of a blend.

In the new joint ventures, PDVSA and its subsidiaries would not continue to give guarantees of any nature as compensation for any action taken by the sovereign state. The state's actions would be its sole responsibility. Moreover, it was specified that the joint ventures would be subject to the production quotas that the state might determine, in exercise of the sovereign rights of the nation in the case of exhaustible and non-renewable natural resources.

12.2.1. Fiscal Stability Guarantees

The anticipated indemnities to be paid by the affiliates to their private partners and, definitively, in the event of expropriation, in the AAs of Petrozuata, Ameriven, and Cerro Negro were based on threshold prices defined by the annual averages of North Sea Brent crude oil – published daily by the commercial press – subject to the implicit price deflator of the USA gross domestic product.

In the case of Petrozuata, Maraven would guarantee the stability of the fiscal regime in full so long as the price of a barrel of Brent was 18.00 US$(1994) or less; the guarantee would be gradually reduced at higher prices, and would reach zero when Brent reached 25.00 US$(1994). But the Ameriven and Cerro Negro AAs would simply guarantee 100% until the price of a barrel of Brent reached 27.00 US$(1996). Maraven's guarantee to Petrozuata was prior to those of Ameriven and Cerro Negro and definitely much more modest (MEM, 03/2006).

However, from 2004 there occurred a strong and unexpected increase in prices. In 2005, Brent averaged 54.44 US$/b. In the case of Petrozuata, the threshold price range was between 22.54 US$(2004)/b and 31.30 US$(2004)/b, while the threshold price for Ameriven and Cerro Negro was 32.51 US$(2004)/b. In response to these high prices, in 2006, the MEM reformed the recent Organic Hydrocarbons Law, so that all hydrocarbon producers had to pay an effective royalty rate of one-third (GOV, 24/05/2006). This legal modification complied with the respective conditions of non-discrimination provided for in the AAs.

Subsequently, the Income Tax Law was also amended so that, as of January 1, 2007, the 50% oil rate was to be applied to all associations (GOV, 09/25/2006). By January 2007, all the AAs were subject to the same tax regime, as regards royalty and income tax. The only pending issue was the majority shareholding in primary upstream activities.

12.2.2. Mandatory Migration

In view of the slowness with which the respective negotiations were progressing, in February 2007, the President of the Republic ordered, by Decree-Law, a mandatory migration within a period of six months. PDVSA would assume a 60% shareholding in both production and upgrading activities.

In the event, only the Sincor association migrated in its entirety. Total, Statoil, and CVP – the latter a partner in all the associations – had, before the migration, shareholdings of 47%, 15% and 38%, respectively. Now, in the new joint venture these would be 30.32%, 9.68% and 60%, respectively. Although the Sincor AA did not specify a threshold price, the agreed guarantees were essentially equivalent to those of Ameriven and Cerro Negro. In the negotiations, the market value for Sincor was estimated to be about five billion dollars, so the CVP had to pay, for an additional 22%, 1.1 billion US$.

However, ExxonMobil, with 41.67% in Cerro Negro, did not agree to negotiate. In accordance with the corresponding clauses of the AA, it preferred to initiate a commercial arbitration against PDVSA/Lagoven before the International Chamber of Commerce in Paris. In December 2011, the arbitration panel in question determined that Exxon was entitled to 908 million US$ in compensation (ICC, 23/12/2011). When this figure is adjusted on the basis of the volumes at issue and the corresponding shareholding interest, it turns out that award was about one-third less than what PDVSA had agreed to pay in the Sincor case. This difference is essentially explained by the quality of the respective upgraders. Sincor built the more complex upgrader, with the extra-heavy crude being upgraded to 31° API; the Cerro Negro Association, on the other hand, built a more modest one, upgrading only to 16° API.

For its part, ConocoPhillips, with its 40% share in Ameriven, likewise did not accept to negotiate and resorted to commercial arbitration before the ICC. In May 2016 an indemnity of 1.5 billion US$, was awarded in its favor (ICC, 24/04/2018). Again, after adjusting for the volumes at issue and the corresponding shareholding interest, this award was about 10% lower than what PDVSA had agreed to pay in the Sincor case. Once again this difference is also explained by the quality of the upgrader: Ameriven was upgrading extra-heavy crude to 25° API.

In these two cases, taking over the shares of the non-migrating parties meant that PDVSA's shareholding already exceeded the 60% threshold, so the one-sixth participation in the Cerro Negro AA of British Petroleum (which had taken over Veba) was not touched, nor was Chevron's 30% participation in the Ameriven AA.

On the other hand, since ConocoPhillips was the only private partner in Petrozuata with a 50.1% shareholding, PDVSA became the sole shareholder in that association. The arbitral tribunal, in the award already mentioned above, determined that Conoco was owed 489 million US$ in compensation, which was a very low figure when account is taken of the volumes at issue and the corresponding shareholding interest, since this figure barely came to a third of what the same tribunal had awarded in the Ameriven case, despite a comparable degree of upgrading being involved. This outcome is explained by the indemnity clause, much more modest in this case.

Had they accepted the proposals advanced at the time in the confidential negotiations with the MEM and PDVSA/CVP, ExxonMobil (XOM) and ConocoPhillips (COP) would both have obtained higher compensation. However, these would necessarily have been final settlements, i.e., ExxonMobil and ConocoPhillips, as part of the settlement, would have had to waive their right to bring international arbitrations against the Bolivarian Republic of Venezuela before ICSID. But this was precisely the intention of both XOM and COP.

12.3. End of Orimulsion

Once Alí Rodríguez took over as President of PDVSA and Rafael Ramírez as Minister of Energy and Mines, they convinced the President of the Republic to put an end to Bitor as an Orimulsion producer. As for Sinovensa, it was allowed to continue operating until December 2006, and thereafter it was to be authorized to produce blends of extra heavy crudes with lighter crudes.

But Bitor had to perform, for a further five years, the Dalhousie (New Brunswick, Canada) supply contract – signed in 1993 –, which provided that, in case of a non-supply of Orimulsion for any reason whatsoever, it would be replaced by residual fuel at an equivalent price. The performance of this contract, in turn, was legally guaranteed by PDVSA's foreign affiliates.

13. Internationalization and Fiscal Control

Between 1983 and 2004, PDVSA had spent 6.85 billion US$ in the acquisition of refining, storage and retail marketing assets, and in equity contributions. Moreover, price discounts on the crude supplies came to another 5.36 million US$ (MEM, 09/2005).

After Minister Rafael Ramírez was appointed to also serve as president of PDVSA in December 2004, the Ministry, under the leadership of Juan Carlos Boué, tried to put an end to these discounts and insisted on the sale of the assets in question. However, this initiative only met with partial success, as Citgo could not be sold due to the intertwining of the power structure around PDVSA. So, the Department of Trade and Supply maintained its control over the export volumes and prices, as well as over coastal shipping operations.

The Ministry also insisted in vain that a strictly national accounting system for PDVSA in Venezuela be reinstated, with a global accounting system to be used only for the operations of PDVSA and its subsidiaries abroad.

14. Domestic Hydrocarbons Market

Oil, as common property, is a free gift of nature for the domestic market (DM), and an international source of rent for the export market (MX). Thus, the prices in question are qualitatively different:

P(DM) = Cost Price +/- {Taxes/Subsidies)

P(MX) = Cost Price + (International Petroleum Ground Rent)

The importance of the domestic market, in terms of volumes, had historically been minor, but it grew in importance from 1970 onwards, after the country's production had reached its peak. Production at the time was 3,707 thousand b/d, of which only 200 thousand b/d – 5.4% – went to cover the domestic consumption of hydrocarbon derivatives. However, by 2006, production was 3,245 thousand b/d, but domestic consumption came to 561 thousand b/d, or 17.3% of the total. The importance of the domestic market, in terms of costs, was even greater since the domestic consumption involved petroleum *products*, whereas an important part of the export basket consisted of *crude oil*.

But the quantitative difference also grew, because domestic prices did not increase at the same rate as export prices. Controlled by the National Executive since the 1930s, domestic prices remained at low levels, but all throughout the concession regime and until the early 1970s, they still covered production costs. But then they began to generate losses.

However, PDVSA did not show these losses in its accounts, and instead only presented the sum total of the domestic and export markets, minus production and refining costs, which obviously produced a positive number. Estimates of costs or losses in the domestic market were never presented. Nor did National Accounts of the BCV differentiate between the domestic and the export market. In spite of the impossibility of finding even a single article dealing with the domestic market that does not refer to it as being subsidized, such subsidies appear nowhere in PDVSA's accounts, nor in the National Accounts of the BCV.

Moreover, the state continued to charge a consumption tax, for example in 2006, of 0.67 US$/b, which formally appears in the BCV's National Accounts as an excise tax on the sector. In short, the price of a barrel of derivatives consumed domestically averaged 7.77 US$. However, as the cost price was 14.64 US$, the real subsidy per barrel was 6.87 US$.

Since the nationalization, the distribution of the IPR among the less advantaged sectors had steadily decreased, so that the low prices of fuels – and therefore, of transportation – became what little that remained of the working class-focused distribution of yesteryear. Hence the social explosion provoked by Carlos Andres Perez when, in 1989, he tried to eliminate these low prices. With Hugo Chávez, subsidies increased again – due to the lag between successive devaluations and increases in domestic prices – and by 2006, they amounted to 1.4 billion US$.

In short, whatever the prices in the domestic market may be, consumers in Venezuela will always have to pay – in one way or another – for the cost of production, but they will never pay for the natural resource as such, since it is a public property. What they might pay over the cost price, does not constitute a ground rent but simply national taxes. Thus, prices in the domestic market are an *eminently* political question.

14.1. Smuggling and Domestic Prices

Low prices in the domestic market also led to smuggling, which grew as the difference between domestic and international prices increased. In 2006, the highest authorities of the MEM and PDVSA estimated it at between 100 and 150 thousand b/d, which implied a loss for the country of between 1.7 billion US$ and 2.6 billion US$. In general, smuggling was always blamed on private individuals transiting the borders with neighboring countries, especially Colombia. However, such large amounts could only be achieved through maritime smuggling. Although bills of lading specified the destination, in coastal shipping for the domestic market there was no control to indicate whether the shipments actually arrived at that destination and in the quantities indicated or whether, while in transit, the destination had been changed or a part of a cargo unloaded. Although the MEM tried to create a control system for coastal shipping, it failed due to resistance from PDVSA's Trade and Supply Department.

Evidently, if the prices in the domestic market were the same as in the international market, smuggling would disappear and with it the corresponding losses. If PDVSA were to act as a withholding agent at the refinery gates or at the ports of import, small numbers of personnel – national guards – would be enough to reduce tax evasion and, ultimately, smuggling to zero. In any event, these were the prices upon which the developed capitalist countries and their international financial institutions insisted. Higher prices would undoubtedly have a negative impact on demand in the domestic market, and that was in the interest of these actors.

Conversely, maintaining the subsidies on hydrocarbon prices meant that taxes in the non-oil sector would have to be increased. Indeed, in the early 1990s, indirect taxes represented less than 5% of GDP(NR); but in the last years of Hugo Chávez's government, they already represented around 15%. Moreover, excise taxes in the domestic market were exceptionally easy and cheap to collect, much cheaper, for example, than value added taxes, administered by the SENIAT. Thus, the increase in domestic prices, in principle, would allow the government to reduce non-oil taxes and even to eliminate the most onerous indirect taxes, in addition to other measures benefiting those most affected by this structural change and, therefore, by the increase in public transport fares.

14.2. Conclusions

Because increasing domestic prices was associated in the public mind with the neoliberal policies promoted by the International Monetary Fund, it was very difficult for any political force to advocate it. In fact, in the discussion on Venezuelan national petroleum policy, no systematic distinction was ever made between oil as a source of ground rent at the international level and as a free gift of nature at the national level. This distinction, evidently, went to the heart of national petroleum policy: the question of territorial property, a question vetoed by modern economic sciences.

In the last years of the Hugo Chávez government, fuel prices at the refinery gate became *negative*, since the regulated prices did not even cover the cost of transportation from the refineries to the gasoline outlets.

15. Sowing the Oil and Public Debt

For Hugo Chávez, the failure of the IV Republic was not due to oil policy, but only to the failure in *sowing the oil*. The nationalization of petroleum and its consequences in the subsequent development policy – indebtedness included –, were not part of his vision of the country. Thus, he would resume the *sowing the oil* and had no doubt that a leftist government, under his presidency, would have the success that the right-wing governments disinterested in the popular welfare and corrupt (from Carlos Andres Perez in his first government, to Rafael Caldera in his second government) would not have had.

Hugo Chávez was sworn in as President at the time of a serious financial and economic crisis. As a first measure to overcome this situation, he recommitted the country to OPEC quotas. Then, contrary to the gloomy perspectives sustained by the top management of the 'old' PDVSA, international prices rose again from 2004, to reach new historic levels. Between 2005 and 2012, the IPR per capita reached the same levels that it had reached between 1975 and 1982. But Chávez also indebted the country again on the basis of these prices, albeit now through indebtedness contracted by PDVSA. Moreover, the People's Republic of China offered him to create a "Heavy Fund", with loans destined to sustain his *sowing the oil*. All in all, by 2013 – Chávez died in March 2013 – although the conventional public debt per capita was the same as it had been in 1998, as a matter of fact it had actually doubled, taking into account the indebtedness of PDVSA and the Chinese Heavy Fund (Figure VI.1).

Figure VI.1

INTERNATIONAL PETROLEUM RENT AND FOREIGN DEBT PER INHABITANTE
1998 – 2013

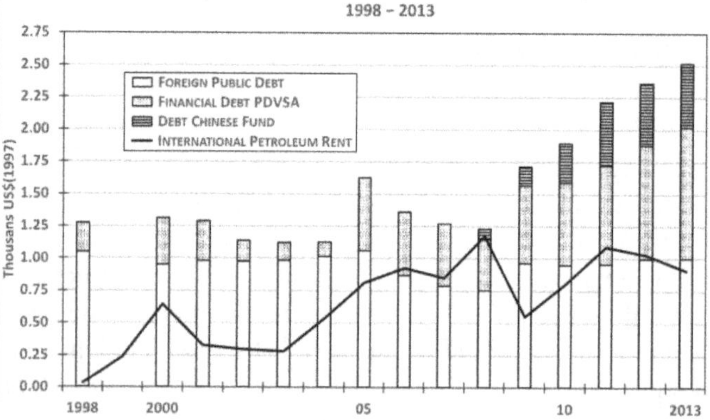

Fuente: José Guerra: *El Legado de Chávez*, 4ª ed. 2013.

Then, when oil collapsed as a source of ground rent in 2014, GDP(NR) per capita followed. Thus, the same cycle was reproduced as during the first government of Carlos Andrés Pérez and the subsequent government of Luis Herrera Campíns. For a start, with massive public spending, in 2006 GDP(NR) per capita surpassed the maximum level reached 28 years earlier, in 1978. It remained at this level until 2014, but plummeted again with falling prices.

Although the Chávez governments distributed most of the IPR among disadvantaged sectors, the investment still exceeded by far the country's capital absorption capacity. Furthermore, human capital had been drastically reduced in the petroleum sector itself after the PDVSA lockdown, and fell even further with the creation of new associations, since the most qualified Venezuelan personnel working with the former operative and association agreement left, in one way or another. Then, the mostly Venezuelan service companies, which had been active for decades around Lake Maracaibo, were also nationalized. Chávez also insisted that the differences in salaries between engineers and less qualified workers be reduced, in accordance with his political discourse around "Socialism of the XXI Century". Based on this same discourse, the space for the private sector in general, particularly in the countryside, was also restricted. As a result, although significant sums were spent on the formation of human capital at the grass roots level, the national managerial capacity was significantly reduced.

Part VII.

Venezuela, an Oil Country: Conclusions

1. Concessionary Regime

At the beginning of the 20th century, with General Cipriano Castro at the head of the government, oil concessions in Venezuela were subject both to mining taxes specified in investment contracts and to national tax sovereignty. However, in 1909, when General Juan Vicente Gómez took over the government, the Mining Law was reformed and the country gave up its tax sovereignty. Furthermore, in the Exposition of Motives of the Law, the natural resource was explicitly considered as a free gift of nature, and mining taxes were to represent only the equivalent of the usual taxes for all other activities.

However, the First World War was to highlight the extraordinary importance of petroleum, not least in terms of future industrial development. Moreover, towards the end of the war, the United States became an importing country (although in retrospect, this was only temporary), and the US-American oil companies turned up in Venezuela in search of concessions. There they were going to compete with the already established European companies, doing so on the basis of practices corresponding to the US-American reality, which was based on the private ownership of the natural resource. In turn, the Venezuelan government then began to consider oil also as a source of international rent and would develop hydrocarbons legislation based on the legislation in the USA.

Thus, in 1920, the first Hydrocarbons Law granted private landowners the preferential right to concessions on their lands, although limited to only one year. The landowners were then to benefit from these concessions by selling them on to foreign oil companies.

However, this legislation would run into criticism and calls for public ownership of the mineral, with a view to the development of the national economy. The state, as the beneficiary of oil as an international source of ground rent, it was argued, should allocate the corresponding revenues to the development of infrastructure, which was practically non-existent, but a *sine qua non* condition for the development of a modern economy. These views prevailed, definitively, with the 1943 Petroleum Reform.

In the 1943 Hydrocarbons Law, the Venezuelan state imposed rents and royalties corresponding to the usual levels in leases on public lands in the United States. The state put itself also on the same footing as the Federal government of the USA, in the sense that contractual relations, as lessor or grantor, would not diminish its sovereign rights. Venezuela thus recovered its tax sovereignty.

However, both the oil companies and the US government officials monitoring the Reform proposed that tax sovereignty be formally limited to the usual level of general taxes in the non-petroleum sector. In other words, tax sovereignty should not be used by the landlord state to increase, at its discretion, ground rent beyond the level defined by the Hydrocarbons Law and the title of the concession. This was the situation prevailing *de facto* regarding the US government. In the United States, private landowners were definitively bound by the original lease, for a term of twenty years, but to last until the exhaustion of the deposits discovered during that term. For its part, the US government never resorted to its tax sovereignty to adjust the lease contracts on public lands.

The USA was definitively not an oil country, whereas Venezuela was one, and would always aspire to a higher international ground rent, precisely by means of its tax sovereignty. However, in the midst of the Second World War, the oil companies and the representatives of the US government had no choice but to accept the position of the Venezuelan government.

For their part, the concessionaires benefited from the Reform by obtaining the renewal of pre-existing concessions *wiping the slate clean*. Moreover, with this renewal, rights previously vested on concession dealers were to remain in force.

1.1. Fifty-Fifty Arrangement

In the Exposition of Motives of the 1943 Hydrocarbons Law, it was stated that the rents and royalties defined by the law would be equivalent, on average and over the useful life of a concession, to a *fifty-fifty*. In addition, the country was to benefit from general taxes which, in essence, were equal to an income tax – recently created, in a process parallel to the Reform – with a rate of 12%. In reality, the profits of the oil companies in Venezuela were subject to the income tax in their countries of origin; and to the extent that the foreign concessionaires paid this tax in Venezuela, they did not have to pay the same amount in their countries of origin, as long as it was lower in Venezuela, which was the case at that 12% rate. Moreover, this was the *pre-war* rate, and US income tax rates during the war were exceptionally high. However, these high levels were maintained after the war, and the Venezuelan government then adjusted the income tax rate accordingly, up to 28.5% for the higher levels of profit. However, due to global price increases, in the cases of the two concessionaires with the best lands, Creole (SONJ) and Shell, the government's share of oil benefits – rents and royalties, *and income tax* – was below 50% in 1946 and 1947.

Then, in 1948, the oil companies proposed to create, *within the Income Tax Law*, an Additional Tax, so that there would be a *fifty-fifty* split of benefits would be achieved, on an *annual* basis, by *adding rents and royalties, and general taxes*. In practice, the additional tax only implied an increase in the effective income tax rate of a few percentage points. Moreover, SONJ and Shell agreed to pay the additional tax *retroactively*, for the years 1946 and 1947, which would allow this to be characterized as a *transaction* that would also commit the state's tax sovereignty.

For its part, the government, regarding petroleum policy, was represented by two outstanding personalities: Rómulo Betancourt and Juan Pablo Pérez Alfonzo. The former was the founder of the then ruling Acción Democrática party, and the latter was the party's spokesman on the matter. Betancourt and Pérez Alfonzo agreed with SONJ and Shell to rewrite history, exalting the importance of the *fifty-fifty*, by means of which a democratic government supposedly achieved what a despotic government had not been able to achieve with the 1943 Petroleum Reform, without ever referring to the issue of tax sovereignty.

Moreover, this *fifty-fifty* was immediately used as a reference by the international oil companies in all the other petroleum exporting countries. Since all of them lacked sovereign rights regarding their concessions, the arrangements were formally of contractual nature, income taxes included.

1.1.1 Additional Tax and Concession Dealers

However, the additional tax hit hard those concession dealers who in the past had secured, with the sale of their concessions, an *overriding royalty*. Until then, they had always paid 12%, since increases to the 28.5% rate only affected the highest incomes. But when the additional tax was introduced, since as landlords they had no costs to deduct, they suddenly found themselves subject to an effective rate of 50%. They appealed to the Supreme Court, claiming an intrinsic value of the mineral resource which should not be subject to an income tax, a position entirely in accordance with Ricardo and, in turn, recognized in USA tax laws. The Supreme Court, without questioning their claims as such, only clarified that they had never been the proprietors of the mineral deposits, as these had always been the sole property of the state.

1.1.2 Landlord State and Tax Sovereignty

Consequently, neither could the royalties of the state be considered simply as a participation in the benefits, as the additional tax assumed. This point of view was taking shape, in 1958, within the ruling civil-military Junta, which included several personalities who had already been in the government of Medina Angarita. The Junta opted not to touch the additional tax, but instead relied upon applying Venezuela's sovereign power of taxation, anchored in the 1943 Petroleum Reform. On the basis of its extraordinary powers, in December 1958, it surprisingly decreed an increase in the maximum income tax rate, taking it from 28.5% to 47.5%. With this, the additional tax to be paid by the concessionaires was effectively reduced to zero, since now the state's participation in the profits was always going to be well over 50%.

1.1.3 Middle East and Intrinsic Value

By limiting itself to an income tax rate of 47.5%, the Junta also had all the other petroleum exporting countries in mind, as this compensated for the existing difference in royalty rates, one-sixth in Venezuela, but one-eighth in the Middle East. In this way, by avoiding any explicit reference to the delicate issue of tax sovereignty, the petroleum exporting countries of the Middle East were going to be able to circumscribe the discussion to the 'intrinsic value' of the natural resource and to *negotiate* the exclusion of the one-eighth royalty from the additional 50% tax, and thus equalize with the Venezuelan tax regime. The marginal rate of the landlord states' government take would then increase, for all of them, from 50% to 56.25%.

1.2. No More Concessions

From its first time in power (1945-1948), Acción Democrática had pronounced itself for a policy of *no more concessions* because it considered that too many concessions had already been granted, and that it was convenient to reserve the remaining promising areas to a national petroleum company, controlled by the state. This led to the creation, in 1960, of the *Corporación Venezolana del Petróleo*, which was to operate on the basis of *assignments* and not concessions. CVP was empowered to promote all types of agreements and to form joint ventures with foreign companies. However, these agreements and joint ventures were to be based on the concession regime. In 1971, CVP signed five such agreements.

1.3. Reversion

Oil concessions had a maximum duration of forty years, extendable for another forty years, upon request from the concessionaires before the 38th year of the term. Therefore, all of the concessions that had been renewed with the 1943 Petroleum Reform had to be extended by 1981 at the latest, or else they would revert to the state by 1983. It was taken for granted that at this point CVP would become a major producing company, with a significant participation in the extended, reverted and renewed concessions.

1.4. Nationalization

However, in 1970, Venezuelan production reached its historical maximum, as a consequence of the depletion of its traditional oilfields. But at the same time and in the context of the events in the Middle East, there was a sharp increase in international prices and, ultimately, in international petroleum rent. At the end of 1973, unexpectedly, the landlord state found itself in an extraordinary position of power.

Faced with this situation, the oil companies made a break for it looking ahead, and promoted the immediate nationalization of their subsidiaries in Venezuela. Once again, Acción Democrática was in government, with an absolute majority in Congress, under the presidency of Carlos Andres Perez. The latter was now alleging that the ultimate aspiration of AD since the 1943 Petroleum Reform – when the party was in opposition – had been to put an end to the concessionary regime, supposedly due to its colonial inspiration.

In effect, with nationalization – more precisely, with the *anticipated reversion* –, the concessionary regime was legally and formally terminated and, consequently, the CVP Service Contracts were also terminated. In this way, a modern concessionary regime, which had nothing to envy when compared to the concessionary regimes of developed European capitalist countries, was abolished to give way to a new *cessionary regime*, with Petróleos de Venezuela S.A., a Holding company, at its head. PDVSA would now act as the regulating entity of petroleum as a productive activity, while the large concessionaires of the past were to become its operating subsidiaries. At the same time, the Ministry of Mines and Hydrocarbons, in its role of regulator of oil as an international source of rent, was to be dismantled.

1.5. Distribution of the International Petroleum Rent and Economic Development

The development of capitalism in an oil-country like Venezuela was inevitably going to give rise to a discussion – between the Right and the Left – as to who should benefit from the fiscal bonanza. Arturo Uslar Pietri became, from 1936, the spokesman of the Right. According to him, this petroleum bonanza had to be channeled for the benefit of the entrepreneurial class as investors, while public spending had to be covered by general taxes from the non-petroleum sector. Formally, these benefits would take the form of loans, and the entrepreneurial class was to give them a useful *purpose*, although, as a practical matter, they would not be repaid. Hence, this was a mechanism to privatize these flows of international ground rent and, consequently, this was part of a process that Marx called the original accumulation of capital, a revolutionary phase of capitalism in which private property is imposed as a *sine qua non* condition for the new mode of production. However, once capitalism was established, private property would become sacrosanct and in principle would only be subject to market exchange.

For their part, the spokesmen of the Left would insist that these revenues should be devoted to the formation of human capital and, to this end, it was necessary to have higher real wages, education, free medical care and other similar benefits, even if in reality, the recipients of these revenues would simply dedicate them to living better.

However, a specific characteristic of rent capitalism is that its development goes hand in hand with a *parallel and continuous process* of original accumulation, that is, of the privatization of a state income, determined by political power. After the Second World War, with the vertiginous increase in production, prices and, therefore, petroleum fiscal income, Betancourt and Pérez Alfonzo came to the conclusion that, by 1948, the national capital absorption capacity of the international petroleum rent had already been surpassed, despite the country having experienced a significant immigration of qualified manpower, as well as the presence of foreign companies in all kinds of activities in the non-petroleum sector. Betancourt and Pérez Alfonzo concluded that exceeding this capacity would bring about higher levels of consumption of imported goods, with overtones of ostentatious consumption, so that for them the solution consisted in lowering petroleum exports accordingly.

1.5.1. Petroleum and Dependence

But this approach was not shared by subsequent governments, nor did it occupy a relevant space in public opinion. Furthermore, the second government of Rómulo Betancourt, with Pérez Alfonzo as Minister of Mines and Hydrocarbons, had to face the drop in international oil prices, and the founding of OPEC in 1960 was assumed as a first step for both the main exporting and importing countries to reach an intergovernmental agreement, a *commodity agreement*, to stabilize prices. This initiative ended in complete failure, with a resounding *no* from the developed countries.

Moreover, in 1966, while AD was still in power, although no longer with a majority in the National Congress, the government attempted to reform the Income Tax Law, resorting to its sovereign power of taxation. On the basis of the approach followed since 1943, summarized in the *Dissenting Vote* of Pérez Alfonzo, which conceptualized the petroleum industry as a public service, the reform was intended to introduce an excessive profits tax. The largest companies, which had the best lands, opposed this measure; instead they imposed their own reform, whereby the maximum rate was simply increased from 47.5% to 52%, and the possibility of tax stability agreements lasting up to five years was introduced. Thus, in short, the companies made sure that Venezuela's tax sovereignty was not going to serve as a reference for the other OPEC member countries (all of which did not enjoy such sovereignty); instead, the situation prevailing in those countries was going to be imposed as a reference for Venezuela.

After the national and international failure of his oil policy, Pérez Alfonzo concentrated on advocating not only for reducing petroleum exports, but also for bringing down the importance of all foreign investments. He then set himself up as the personification of the landlord state and, as such, finally assumed a Ricardian position, attributing an intrinsic value to *exported* oil. However, he did not assume this position vis-à-vis the concessionaires, but rather vis-à-vis Venezuelans, who were therefore supposed to put the IPR to a use that would contribute to the development of the country, and thereby reduce dependence on it. Only in this way could the international petroleum rent be legitimized. Conversely, for Pérez Alfonzo, simply consuming the international petroleum rent was illegitimate and, ultimately, an act of corruption.

Subsequently, with production falling and prices rising, Perez Alfonzo would insist on even greater export cuts: the higher the prices, the lower the volumes.

1.5.2. Great Venezuela

In contrast, Carlos Andrés Pérez attributed the failures of the past to the economic policy of the time – import substitution – and decided to take a great leap forward on the basis of a new economic policy, with the importation of capital goods. To this end, he resorted not only to the high IPR of the present, but also that of the future, i.e., he indebted the country.

This was a noteworthy fact, given that the country had freed itself from any international debt in 1930, taking advantage of the fiscal bonanza due to oil. Moreover, the oil companies were very interested in preserving this situation, since they were aware that, in any fiscal emergency, the governments in power would resort to them in the first place. But now, with nationalization, indebtedness would become a lever to reopen the country to international oil capital.

Both Carlos Andrés Pérez and Pérez Alfonzo put forward *national projects*, although they were diametrically opposed to each other. In either case, both projects would have required a broad national agreement. However, both leaders – and also Betancourt – maintained their sectarian position on petroleum policy. According to this view, nationalization had supposedly not been the result of a successful national petroleum policy, a historical product of the most varied political regimes as Rafael Caldera pointed out; but solely the result of the nationalist policy of Acción Democrática, with its supposed roots in the *Dissenting Vote* of 1943.

2. Cessionary Regime

Now, PDVSA and its subsidiaries were supposed to liberate the country, as Betancourt said at the time, 'from the presence on its soil of aggressive foreign economic and political forces'. It crossed no one's mind that there might be a need to analyze how the nationalization of the foreign concessionaires – undoubtedly an event of capital importance – was going to affect the institutional, economic, political and social development of the country, as well as the *checks and balances* of a liberal democratic regime. There was definitively no room for doubt that, with nationalization, the country was going to be better off.

On the other hand, references to the limited capital absorption capacity entirely disappeared from public discussion. The collapse of the national economy would only be explained by the incapacity of the governments in power, which were blamed for having insisted, demagogically, on a distribution of the international petroleum rent focused on the lower classes. To reinforce this explanation, it became indispensable to resort to the phantom of corruption. Following this line of thought, the country's indebtedness was simply a consequence of generalized misgovernment.

2.1. Nationalization and Petroleum Policy

There was not even an appreciation of the fact that, with the nationalization, the country had transitioned from one petroleum regime to a different one and, consequently, from one petroleum policy to another. Nationalization was understood as the end of history when, in fact, it deepened the historical fault lines in the country. A state is defined by the control it exercises over a certain territory and over all the natural resources it may contain, but now PDVSA appropriated the most economically important natural resource, and was going to manage it, as a productive activity, at its convenience, without paying any heed to the country as the owner of the natural resource.

Trained and educated by the foreign concessionaires for decades, the upper management of PDVSA and its subsidiaries had assimilated all their viewpoints and resistance to any advance by the landlord state. The successes of the landlord state were attributed to good luck, circumstantial events and the like, but never to a successful policy. However, this interpretation of the history of the concessionary regime was facilitated by the way in which the landlord state itself presented its own progressive steps. All of them were supposed to be only about justice, bringing to an end – although always partially – situations that were unjust from the start, imposed by the foreign concessionaires. For example, the concept of reversion of concessions was supposedly grounded on provisions from the early twentieth century, and not to the actual evolution of the country and the necessary change in the corresponding policy. Some went so far as to argue that reversion meant recovering a granted *activity*, just as in the case of an electricity concession (Pérez Alfonzo, 1970). Likewise, what happened throughout the 1960s had only been a defense of fiscal income per barrel, and an attempt to impose a 'fair' price, with the price in the USA domestic mark considered as such. In other words, at the political and economic levels, it was never acknowledged that Venezuela and OPEC imposed a proprietary rent against lower prices.

Furthermore, any textbook of Economic Sciences would stress that the optimal ground rent would be the result of free competition among the corresponding producing companies. The large foreign concessionaires, on the other hand, were always accused before the national public opinion of arranging, by means of the most diverse machinations, not to contribute to the national treasury with their fair share. How could there be any doubt: this could never happen with PDVSA!

What was not understood was that the exporting countries had been confronting foreign consumers through the concessionaires, and that now Venezuela was to confront foreign consumers through the state-owned national petroleum company. PDVSA would again put forward, one by one, the same arguments as the concessionaires. These arguments had been *politically* defeated in the past because they had come from foreign concessionaires. Now, because they were coming from the national state petroleum company, these same arguments were going to be imposed with relative ease. Thus, to observe the development of petroleum policy during the cessionary regime is akin to watching the video about the history of the concessionary regime being rewound. The first step was for PDVSA and its subsidiaries to oppose OPEC quotas. Subsequently, with the Bolivarian Constitution of 1999, PDVSA, the holding company, formally became the regulator of oil as a productive activity; the Ministry had already abandoned its role as the regulator of petroleum as a source of international ground rent. The last step consisted in renouncing the exclusive national jurisdiction in this matter, and subjecting the Nation to international arbitration, a space dominated by developed capitalist countries, something that would have been simply impossible under the concessionary regime.

Of course, there was no lack of critical voices at all levels in this process, which reflected the memory of how successful the concessionary regime had been, precisely by creating a proprietary fiscal regime. However, although the arguments put forward by those critical voices were pertinent and solidly based, the reality was that the national ownership of the natural resource as an international source of ground rent no longer had an institutional seat, and therefore critical arguments were accorded nothing more than fleeting momentary glimpses.

2.2. Petroleum in Venezuelan History

This situation created by nationalization was reflected, of course, in all publications on the subject. As an example, let us look at some publications of the highest academic level.

In 1997, Fernando Coronil published his book *The Magical State* in which he presents a historical vision of both the Venezuelan 20th century and the failure of the landlord state. Regarding the failure of Carlos Andrés Pérez and his economic policy, the author relies on his own research and that of Julie Skurski, particularly with respect to the policy of creating an automobile industry and a tractor industry, and also analyses a prominent case of corruption. In this book, however, international oil companies do not appear as actors; and Coronil does not mention PDVSA and its subsidiaries even once.

In 2000, Asdrúbal Baptista edited a three-volume work, *Venezuela Siglo XX: Versiones y Testimonios*, which included sixty contributions from leading professionals on a wide variety of topics. Among them were four historians: Manuel Caballero, Germán Carrera Damas, Elías Pino Iturrieta and Ramón J. Velázquez. Inevitably, they were all going to mention the impact of petroleum on 20th century Venezuela, but not one of them even mentioned the 1943 Petroleum Reform, nor did they attribute any importance to the 1975 nationalization, much less to PDVSA and its subsidiaries.

In 2010, Mirtha Rivero published *La Rebelión de los Náufragos*, based on interviews with dozens of personalities of the time, in an effort to clarify the reasons for the fall of Carlos Andrés Pérez in 1993. In the statements of the interviewees, any reference to petroleum policy and, therefore, to PDVSA and its subsidiaries, is conspicuous by its absence among the innumerable problems Pérez had to face as President.

Once the foreign concessionaires were nationalized, the problem of how the country was going to manage petroleum as a source of an international ground rent, on the one hand, and as a productive activity, on the other, simply became unimportant. Moreover, this was what mainstream Economic Sciences asserted; the state could – and indeed should – limit itself to collecting extraordinary profits.

2.3. Distribution of IPR, Devaluations and Non-Petroleum Exports

During the first administration of Carlos Andrés Pérez, massive investment in the development of the Great Venezuela brought about an increase in real wages. However, in 1983, the inevitable failure of his economic policy led to the devaluation of the bolivar and the loss for good of its historical stability. Now, investments, together with all types of financing – in addition to international indebtedness – were generally based on formal contractual relations denominated in dollars, and the government would grant them a preferential exchange rate. But salaries and all types of social benefits, which represented IPR distributed to the lower classes, were denominated in bolivars and were only going to be adjusted belatedly and partially, reflecting the devaluation of the national currency, or simply not adjusted at all.

The distribution of IPR focusing on the lower classes was coming to an end and, faced with a shrinking domestic market, economic policy would focus on promoting non-petroleum exports which, in order to be internationally competitive, would require the lowest possible real wages. The public administration would have to set an example, and public expenditure would have to be covered by an increase in taxes in the non-oil sector, in line with what Arturo Uslar Pietri had been arguing since 1936.

In the national press it was stated, with great pomp, that Venezuela was becoming a post-rentier country, whereas a specific policy regarding petroleum as a source of international ground rent, which had been the essence of AD and Copei's petroleum policy in the past, was now simply considered irrelevant. When the price collapse occurred in 1998, PDVSA was still declaring to the four winds that it would not reduce its production by even a single barrel. This would bring about the end of AD and Copei, and with them the collapse of the political regime that had emerged in 1958 with the Pact of Punto Fijo.

2.4. Hugo Chávez and the V Republic

There is no doubt that with the electoral victory of Hugo Chávez there occurred a radical rupture in the political history of the country. However, Chávez did not relate the crisis of the country to the petroleum policy that had arisen with nationalization. His petroleum policy would be different from the previous one only in terms of a defense of OPEC quotas, although extra-heavy crude from the Orinoco Oil Belt, processed into Orimulsion or in upgraded crudes, would continue to be excluded. Thus, in any case, OPEC would continue to loose importance for Venezuela.

Moreover, Chávez did not understand either the relationship between the Oil Opening policy and the country's foreign indebtedness. Certainly, besides becoming subsidiaries of PDVSA, the operating companies were handed over to the upper management of the former concessionaires, and there is no doubt that, from the beginning, the Internationalization policy was just the first phase of what was later to be the Oil Opening. However, PDVSA and its subsidiaries were only able to impose their position once the International Monetary Fund and other international financial institutions had conditioned their credits accordingly. Obviously, Venezuela, as an oil country, should have been a creditor country and should never have become a debtor country. This is the case in the world of the developed countries, of Alaska and Norway, which recognized the limitations of their capital absorption capacity.

However, Venezuela was no longer perceived as a petroleum exporting country, but as a capital importing country. Thus, when the opportunity arose, the Chávez government returned to the distribution of IPR, focusing on the lower classes and embarking on large investment projects, without any consideration of the availability of human capital, which during his government was still shrinking. Also, instead of reducing international public debt, Chávez continued indebting the country based on peak prices. He did not take PDVSA off the rails upon which the 1975 nationalization had placed it, although, for the time being, the privatization of productive activities was stopped through the renegotiation of the operating and association agreements of the 1990s. However, inevitably, prices would go down again and PDVSA would once again move forward along the same tracks. At the end of the years of Hugo Chávez as President of the Republic, PDVSA still retained its role as the institutional entity managing the natural resource in terms of petroleum as an essentially privatized productive activity, with the state as the direct guarantor of the profits of the operating companies, within an international legal arbitration framework.

Obviously, this whole process, from the creation of PDVSA to the Oil Opening, was not only questionable from the point of view of national law, but actually incompatible with it. However, there was no formal involvement in this process on the part of the international oil companies which, 'in good faith', would have 'trusted' the representations made by PDVSA and the National Congress. Thus, their 'legitimate expectations' were to become the starting point of any international arbitration.

3. Conclusions

The policy of using petroleum as an international source of ground rent is essentially *governmental* and, above all, *national*, in the same sense by which the country sees the *national* armed forces as being in charge of the integrity of the national territory.

The success of petroleum policy as a source of international ground rent requires, first and foremost, *national unity.* This unity was broken with the *fifty-fifty* split of 1948 and, from then on, the concessionaires were going to take advantage of the rivalry between two domestic factions. Subsequently, the division of the country deepened with nationalization. In reality, PDVSA never convinced the country; it imposed its views in alliance with the international oil companies and financial institutions, backed by the consumer countries. Despite this, the country would continue to view PDVSA as the pride of the nation.

As regards the policy of national distribution of international petroleum rent, this is a political decision of the Left or of the Right, in terms of whether it is more favorable to labor or to capital (although, obviously, this policy requires compromises without which the required national unity cannot be maintained). However, the division of the country would go on deepening over time, once the Project of the Great Venezuela had failed, under both Rafael Caldera and Hugo Chávez. Nevertheless, in spite of extreme changes in circumstances, the *Oil Opening*, with its project of a 'post rentier' country, was to continue moving ahead.

Part VIII.

Arbitrated Sovereignty

The subordination of land tenure to capital is inseparable from the development of capitalism. In the prevailing economic thought, this subordination manifests itself in the disappearance of natural resources, i.e., land, as a factor of production, and their absorption by capital. Economic science only conceives ground rent as part of the extraordinary profit of capital and, as such, it is seen solely as representing a differential rent. Now, when empirical phenomena contradict economic theory, as occurs, for example, with the typical embodiment of mining rent under conditions of private mineral property, that is, the royalty, economic science takes refuge by resorting to justifications that consider mineral deposits as accumulations of valuable 'natural' commodities.

However, in the case of the development of capitalism at a national level, if the minerals in question were destined for the domestic market, then the ideal solution was public ownership and a concessionary regime. In this case, whatever the state collects by way of levies will always be taxes and not ground rent (even if the levy is called a royalty), so this income will become part of the national budget, like any other tax. On the other hand, when it comes to exports, whatever the state may collect beyond the usual profit on the invested capital, is an international ground rent, even if it is called income tax.

1. Foreign Investment and Sovereignty

Developed capitalist countries consider that, in principle, foreign investments should have the same rights as domestic investments. Therefore, in case of disputes, foreign investors had to resort to national courts. However, once these instances had been exhausted, the foreign investor still had the option of resorting to the government of his country to espouse his cause (literally, to stand together bound by handcuffs). In such a case, his government would approach the host government with a view to resolving the dispute diplomatically, for example, by submitting the dispute to international arbitration.

The Latin American countries, colonized in the proper sense of the word mainly by Spain and Portugal, preserved the corresponding legal regimes upon independence (from the beginning of the 19th century), despite their economic, political, and social underdevelopment. However, towards the end of the 19th century, these countries thought it necessary to shield themselves against procedures that could involve the international sphere when it came to sovereign decisions regarding natural resources, such as, for example, the granting of mining concessions. They then made use of what became known as the Calvo Doctrine. Thus, disputes over mining concessions were to be stringently restricted to the national legal sphere. This was also the case in Venezuela.

2. Permanent Sovereignty over Natural Resources

On the other hand, in the underdeveloped countries of the Middle East, Africa and Asia –which were dominated throughout the 20th century by the European imperial powers and, finally, also by the USA –, mining concessions had colonial or semi-colonial roots, which is why the sovereign rights of such countries were denied as a general matter, and not only in terms of their rights relating to tax sovereignty. These rights would be adjudicated in arbitration proceedings, based on the principles of law common to the parties involved in disputes, and on 'the general principles of the law of civilized nations', i.e. of developed capitalist countries. Even when the countries in question had already acceded to their independence, their mining concessions would still be subject to international arbitration, and the countries that owned the natural resources involved were still being denied their tax sovereignty.

However, in the early 1950s, the United Nations (UN), with a growing membership of newly independent countries, brought this situation into question, and, in 1952, the General Assembly approved a resolution on the "Right to exploit freely natural wealth and resources" (UN, 21/12/1952, Resolution 626). Its preamble stated that "the right of peoples freely to use and exploit their natural wealth and resources is inherent in their sovereignty and is in accordance with the Purposes and Principles of the Charter of the United Nations". It was recommended that countries make use of these rights "wherever deemed desirable by them for their own progress and economic development" and warned "all Member States to refrain from acts, direct or indirect, designed to impede the exercise of the sovereignty of any state over its natural resources". By then, the confrontation between the British and Iranian governments over the nationalization of AIOC was already in full swing.

In 1962, the General Assembly adopted another resolution on the subject, "Permanent Sovereignty over Natural Resources" (UN, 14/12/1962, Resolution 1803). Its declaration of principles included:

> Nationalization, expropriation or requisitioning shall be based on grounds or reasons of public utility, security or the national interest which are recognized as overriding purely individual or private interests, both domestic and foreign. In such cases the owner shall be paid appropriate compensation, in accordance with the rules in force in the State taking such measures in the exercise of its sovereignty and in accordance with international law. In any case where the question of compensation gives rise to a controversy, the national jurisdiction of the State taking such measures shall be exhausted. However, upon agreement by sovereign States and other parties concerned, settlement of the dispute should be made through arbitration or international adjudication.

One of the Recitals expressly referred to the aforementioned principle, stating that this Resolution did not provide for anything which

> [...] in any way prejudices the position of any Member State on any aspect of the question of the rights and obligations of successor States and Governments in respect of property acquired before the accession to complete sovereignty of countries formerly under colonial rule.

In reality, 'permanent sovereignty' consisted of the right of newly independent countries to revisit existing mining concession arrangements, renegotiate them, or simply change them by force of law at the risk of confrontation.

Finally, in 1966, the United Nations returned once again to the subject, insisting, in particular, on the fact that:

> [...] natural resources are limited and in many cases exhaustible and that their proper exploitation determines the conditions of the development of the developing countries both at present and in the future (Resolution 2158).

2.1. Organization of the Petroleum Exporting Countries

Until 1966, OPEC never referred to the sovereign rights of member countries, much less to their tax sovereignty. During the fifty-fifty negotiations in the Persian Gulf countries, the fiction was always maintained that it was a matter of remedying a conceptual error, because the royalty appeared, for income tax purposes, first as a cost of production and then, in the total sum, as an integral part of the *fifty-fifty*, without taking into consideration the 'intrinsic value' of the mineral.

By 1968, the situation was already different. In the preamble of the *Declaratory Statement of Petroleum Policy in Member Countries* (Resolution XVI.90), OPEC insisted on "the inalienable right of all countries to exercise permanent sovereignty over their natural resources in the interest of their national development is a universally recognized principle of public law and has been repeatedly reaffirmed by the General Assembly of the United Nations, most notably in its Resolution 2158 of November 25, 1966".

The Declaratory Statement then referred specifically to the *Settlement of Disputes*:

> Except as otherwise provided for in the legal system of a Member Country, all disputes arising between the Government and operators shall fall exclusively within the jurisdiction of the competent national courts or the specialized regional courts, as and when established.

Some member countries, with very weak legal structures, could then resort to regional courts – yet to be created –, which would replace the existing provisions on international arbitration. Such institutions would be based on the national laws of exporting countries with more developed legal regimes.

2.2. Petroleum Nationalizations and the New International Economic Order

Resolution XVI.90 stipulated that governments would be entitled to a "reasonable" participation in the share capital of the old concessions, "on the grounds of the principle of changing circumstances". This would be a policy of participation, as opposed to a policy of nationalization, and thus OPEC could finally achieve what had been its main purpose when it was founded: to rescue the structure of international production control based on the US domestic market and, with it, the corresponding price structure.

While the negotiations were initially successful, events precipitated after the Third Arab-Israeli war of 1973. The international concessionaires would then withdraw by advancing, in the form of effectively promoting their own immediate nationalization. This took place *de facto* in 1974, although *de jure* it was to be implemented over a couple of years. Saudi Arabia, Iran, Iraq, Iraq, Kuwait and Venezuela were to nationalize in full; in the other member countries, state interests acquired majority shareholdings. The international concessionary regime was thus brought to an end.

The euphoria that took hold of the exporting countries on seeing their international petroleum rents multiply was shared, in political terms, by the Third World in general, as the process of decolonization also culminated in these same years. In particular, this euphoria was expressed in the *United Nations Declaration for the Establishment of a New International Economic Order* of May 1, 1974.

In reality, 'permanent sovereignty' had also been brought to an end. Now all oil-producing countries were to be simply sovereign, just like Venezuela, with control over prices and volumes at their own discretion and, in the midst of the euphoria, they lost sight of an issue as essential as *Dispute Settlements*.

However, the developed capitalist countries were aware that the oil countries would now have to control national, state-owned, oil companies, and to manage the control of international production, tasks for which they lacked the requisite political, social and institutional structures. With regard to the latter, in particular, such control could only be attempted on the basis of a compromise put together by the developed countries – capitalists and consumers – and their international oil companies, and accepted by the exporting countries and their landlord states. But the developed capitalist countries were unwilling to compromise.

3. International Investment Arbitration

3.1. International Center for Settlement of Investment Disputes

When the world economic crisis of 1929 broke out, the governments of the developed capitalist countries, which already had significant public budgets, took counterproductive measures to balance budgets at the national level by cutting public spending, when what would have been appropriate would have been deficit spending; at an international level, each country tried to save itself by adopting protectionist policies.

However, in 1944, at the Bretton Woods Conference, with the defeat of Germany and Japan already foreseeable, the countries that would end up as victors of the Second World War, under the leadership of the United States, agreed to create the World Bank and the International Monetary Fund. These two international financial organizations, headquartered in Washington, were to be responsible for the international coordination of the macroeconomic policies of national governments in the post-war period. They would also be responsible for promoting the investment of public and private capital throughout the world, with a view to the economic development of the most backward countries.

The Soviet Union, which played a major role in the defeat of Germany, also participated in this Conference, although it would take no part in the Agreement. The end of the Second World War led almost immediately to the Cold War. For the developed capitalist countries, the question would become one of ensuring that the already unstoppable process of decolonization would lead to the globalization of capitalism, and not to the advance of communism.

A central concern of capital exporting countries – a very small number of European countries and the USA – was the security of the investments of their nationals in capital importing countries. In practice, all existing arrangements – in case any disputes arose – required the involvement of the two governments in question. The home country of the investor would have to espouse the investor's cause *after* a dispute had arisen and, if necessary, pressure the host country government to reach some settlement, possibly through international arbitration. Then, in 1965, the World Bank proposed to create within itself an international arbitration tribunal for the settlement of differences and disputes between foreign investors and host states. Thus was born the *International Centre for Settlement of Investment Disputes* (ICSID), conceived as an international treaty. It was to enter into force in 1966, having been ratified by twenty countries. Essentially, this treaty was to be based on the premise that, before a foreign investment was made in a given country, the investor and the state in question had to reach an investment agreement whereby the beneficiary state would definitively and irrevocably agree to submit to arbitration before the ICSID if the investor so wished[1].

Subsequently, ICSID suggested that such consent to international arbitration should form part of bilateral and multilateral investment treaties between states. Thus, when a dispute arose, all that a foreign investor from one of the contracting states would have to do was to send a letter to the

[1] On the role of major international oil companies in the creation of ICSID, see Boué, 2021.

government of the host state accepting the latter's unconditional offer to go to international arbitration.

The arbitral award would emanate from an instance superior in the judicial hierarchy to the highest court of the respective Contracting States and, indeed, from the highest court of any ICSID member country. All such member countries unconditionally guaranteed by treaty the enforcement of the corresponding awards.

Arbitral tribunals were to be composed of three arbitrators, whose nationalities would be different from those of the disputing parties. One would be appointed by the claimant: the private investor; the second would be appointed by the respondent: the state; and the third, as president of the tribunal, would be appointed by mutual agreement. But in the event that no agreement was reached, the presiding judge would be appointed by the President of the World Bank. Such arbitral tribunals would have, as the German phrase goes, Kompetenz-Kompetenz as to all matters relating to jurisdiction under the treaties. For example, if a respondent state has objections as to the competence of ICSID to deal with a specific case, for whatever reason, the decision thereon would be taken by the same tribunal adjudicating on the dispute.

ICSID arbitral awards were to be final, i.e., there would be no second instance. Furthermore, arbitral awards would not create binding precedents. This means that two tribunals acting in essentially identical circumstances may reach opposite conclusions. There only exists the possibility that one of the parties may move for the annulment of the award on formal grounds, such as the incorrect constitution of the tribunal, or the fact that the tribunal has made decisions without justification. Decisions on annulment are taken by an Annulment Panel composed of three members, all appointed by the President of the World Bank.

Hence, a truly global Supreme Court, placed above all national Supreme Courts, in the form of an arbitral tribunal based on bilateral and multilateral investment treaties, arose from the World Bank. In this global Supreme Court, the claimants are the private investors, and the defendants are the nation states. If necessary, the presiding judge is appointed by the President of the World Bank.

3.2. Historical Development

ICSID, as an international treaty, entered into force in 1966, after having been ratified by five developed capitalist countries (all exporters of capital) and fifteen African countries belonging to the British Commonwealth. However, the continental Latin American countries rejected it.

Also in 1966, the United Nations established the Commission on International Trade Law (UN, 17/12/1966, Resolution 2205). In 1976, it presented its rules on commercial arbitration, which were approved by the United Nations (UN, 15/12/1976, Resolution 31/98). Thus, *ad hoc* arbitration or other types of arbitration not covered by ICISD would have somewhere to look for legal redress.

For its part, the Soviet Union was not going to participate in the Washington-based ICSID, which it considered a US-dominated institution. However, in 1970, the Soviet Union agreed with the United States to resort to the Arbitration Institute of the Stockholm Chamber of Commerce in case of East-West trade disputes. This Institute had a long history in the field dating back to 1917.

ICSID's early development was very slow as far as both membership and arbitration cases were concerned. However, after the collapse of oil prices in 1986, coupled with the collapse of the Soviet Union from 1989 onwards, ICSID's development accelerated. The stream of investment treaties with incorporated international arbitration clauses became a torrent. Today, there are more than 2,500 bilateral investment treaties.

The first among these treaties were generally signed between developed and underdeveloped countries. As for petroleum exporting countries, Indonesia and Nigeria became members of ICSID in 1968, Ecuador in 1986 and Venezuela in 1994. At present, only two OPEC member countries do not belong to ICSID: Libya and Iran.

Subsequently, a large number of bilateral investment treaties were signed between OECD member countries, on the one hand, and the newly independent former Soviet republics and Eastern European countries, on the other. Moreover, the European Union, with its eye on the hydrocarbon wealth of these areas, took the initiative to negotiate a multilateral investment treaty. The first result, in December 1991, was the European Energy Charter, a non-binding declaration of principles, which then served as the starting point for the subsequent Energy Charter Treaty (ECT) negotiations, concluded in December 1994. For now, let us move on to Venezuela.

4. Venezuela-Netherlands Bilateral Investment Treaty

In the "Agreement on encouragement and reciprocal protection of investments between the Kingdom of the Netherlands and the Republic of Venezuela", Dutch investors in Venezuela were to be granted the same rights as Venezuelan investors in Holland. Its first article listed the investments it was to cover, which included:

> [...] rights granted under public law, including rights to prospect, explore, extract, and win natural resources.

This was the only time this treaty would mention natural resources, and it did so only from the perspective of investors. Conspicuously absent from the picture was the notion of a sovereign state in charge as such of the national territory and all its components, with aspirations to capture an international ground rent. Thus, even as regards this most elementary of all sovereign acts was concerned, the treaty provided:

> Disputes between one Contracting Party and a national of the other Contracting Party concerning an obligation of the former under this Agreement in relation to an investment of the latter, shall at the request of the national concerned be submitted to the [ICSID], for settlement by arbitration or conciliation under the Convention on the Settlement of Investment Disputes between States and Nationals of other States [...].

And in order to comply with ICSID requirements, the treaty specified that:

> Each Contracting Party hereby gives its unconditional consent to the submission of disputes [...] to international arbitration.

According to this treaty, Dutch investors in Venezuela, and Venezuelan investors in the Netherlands, are defined as such by their respective national law. In Venezuela, the foreign investor only has to comply with an elementary bureaucratic process, identifying himself as a Dutch investor; in Holland, as indicated above, all that is required to qualify as such is a post office box and the services of a law firm that takes care of complying with the minimum legal requirements, without having to have even a single employee in the country. In Venezuela, both ExxonMobil (XOM) and ConocoPhillips (COP) were going to appear under the guise of Dutch investors when the time came.

The corresponding investments in Venezuela were not originally undertaken as Dutch investments, since XOM and COP considered themselves sufficiently protected by PDVSA's contractual guarantees, based on the corresponding threshold prices and subject to international arbitration before the International Chamber of Commerce in Paris. But, as of 2005, when prices reached unexpected levels and it was already obvious that the Chávez government was going to increase the fiscal participation in the existing associations, both XOM and COPP incorporated the Netherlands into their respective corporate chains of ownership, in order to be able to resort to ICSID arbitration based on the bilateral investment treaty in question.

Any arbitral award rendered under the Venezuela-Netherlands BIT, would be based on:

> i. the law of the Contracting Party concerned;
>
> ii. the provisions of this Agreement [...] between the Contracting Parties;
>
> iii. the provisions of special agreements relating to the investments;
>
> iv. the general principles of international law [...].

Each Party pledged not to take "any measures to expropriate or nationalize investments of nationals of the other Contracting Party or take measures having an effect equivalent to nationalization or expropriation with regard to such investments", except pursuant to the payment of "just compensation". However, if the parties were to fail to reach a negotiated agreement on the amount of such compensation, this would be determined by arbitration:

> Such compensation shall represent the market value of the investments affected immediately before the measures were taken or the impending measures became public knowledge, whichever is the earlier, it shall include interest at a normal commercial rate until the date of payment [...].

In what follows, we will examine this issue by concentrating, by way of example, on the XOM and COP extra-heavy crude oil upgrading associations, since, in 2007, both would resort to ICSID proceedings. We will leave aside the *Exploration at Risk and Profit Sharing associations* of XOM and COP, i.e. the La Ceiba and Corocoro AAs, which are of little importance.

4.1. ExxonMobil

In the Cerro Negro AA, both XOM and CVP had a 41⅔% shareholding, and British Petroleum the remaining 16⅔%; Exxon had merged with Mobil in 1999, and British Petroleum had acquired Veba in 2001. In order for CVP to reach a 60% share in the new Joint Venture, according to the mandatory migration policy that had to be completed by June 26, 2007, XOM's and British Petroleum's shareholdings had to be reduced, respectively, by 13.1 and 5.2 percentage points. But with XOM refusing to migrate and consequently being expropriated, British Petroleum was to keep its 16⅔%, and the state was to compensate XOM for its 41⅔% share in the Association. Negotiations began, although in September 2007, XOM initiated the formalities required to trigger arbitration against the Venezuelan state before ICSID.

4.1.1. High Court of Justice in London (XOM vs. PDVSA)

However, in January 2008, XOM surprisingly resorted to the High Court of Justice in London. It requested the freezing of all of PDVSA's assets outside Venezuela for up to US$ 12 billion, as a precautionary measure in connection with the suit it was about to file before the I against PDVSA. Such a measure was only admissible because it involved a claim between private parties, and PDVSA, as a corporation, was a private company, even though the state was its sole shareholder. On the other hand, ICSID does not allow precautionary measures to be taken against states, so the private investor can only resort to compulsory measures when it is a matter of enforcing a final arbitration award. In fact, this was XOM's main reason for resorting initially to the ICC, although it was going to continue with its arbitration against the BRV before ICSID.

The High Court granted the measure, requested *ex parte* – i.e., without prior notice to PDVSA – and, therefore, on a provisional basis. This kind of procedure is justified, for example, in the case of a suspected criminal entity and its bank account, as any prior notice would give it the opportunity to empty the account in question. But this was not the case with PDVSA and its assets abroad. In reality, as far as XOM was concerned, the idea was to bring PDVSA, and thus the Bolivarian Republic of Venezuela, to its knees. However, in subsequent hearings, the High Court concluded that there was no valid cause for such a procedure, and lifted the freeze in March 2008. Furthermore, the failure of an *ex parte* measure meant that XOM had to pay all the costs incurred by PDVSA in its defense (High Court of Justice, 20/03/2008, Folio 61).

4.1.2. International Chamber of Commerce (XOM vs. PDVSA)

In April 2008, XOM filed its request for arbitration against PDVSA with the ICC, pursuant to the latter's assurances in the AA regarding 'discriminatory measures'. More specifically, XOM considered as 'discriminatory' the increase of the royalty rate from 1% to 16⅔%, in October 2004; the introduction of the extraction tax of 33.33% pursuant to the reform of the Organic Hydrocarbons Law, in May 2006; the increase of the income tax from 34% to 50%, pursuant to the reform of the Income Tax Law of September 2006; and the imposition of production and export quotas in 2006 and 2007. Finally – and most importantly – XOM was, of course, entitled to compensation for its expropriation which, according to XOM, should be valued at 6.5 billion US$ (ICC XOM vs. PDVSA, 23/12/2011, §99).

The Tribunal concluded that only the increase of the income tax from 34% to 50%, and the imposition of production and export quotas in 2006 and 2007, qualified as discriminatory measures.

In order to estimate the compensation to be paid for the expropriation, the Tribunal was to take actual data for the year 2007 as the basis of calculation. However, for the period 2008-2035 (when the term of the AA would have ended), the Tribunal would have had to build up a production profile and incorporate the corresponding costs, and also subject the forecast prices to the threshold defined by the AA, in order to estimate the net compensation – after applying the income tax rate of 34% – corresponding to XOM.

Then, to determine the net present value – the market value of the AA – a discount rate was to be applied, but the AA did not specify anything in this regard. This was a practical rather than a legal issue, although one of great importance. In the negotiations leading to the association, XOM – just like all the other companies – had always insisted on the high risk of the project and, consequently, on the corresponding contractual conditions. In contrast, in its submissions to the tribunal, XOM now minimized the risks, so that the applicable discount rate would be rather low. For its part, PDVSA stressed the high risks of all kinds, typical of this kind of project. To give the reader an idea, at a discount rate of 6%, reflecting a low risk, the net present value of 100 US$ payable within ten years is 55.84 US$; but at a discount rate of 18%, reflecting a high risk, the net present value is only 19.11 US$. The tribunal accepted PDVSA's position and set the discount rate at 18%. Finally, the arbitration award, in December 2011, established a compensation of 908 million US$, which PDVSA paid immediately.

4.1.3. International Center for Settlement of Investment Disputes (XOM vs. BRV)

In the meantime, XOM and BRV each appointed an arbitrator before ICSID. Due to the lack of agreement between the Parties, the President of the World Bank appointed the President of the Tribunal (ICSID XOM vs. BRV, 2014).

4.1.3.1 Jurisdiction

The first issue to be decided was ICSID jurisdiction. XOM had only become a Dutch investor on February 21, 2006. The Tribunal would only have jurisdiction over disputes arising after that date, so it had no jurisdiction over the increase in the income tax rate from 34% to 50%, since the possibility of such an increase had been in dispute prior to the date indicated (ICSID XOM vs. BRV, 2014, §15). This was, simply and definitively, a fact. The Tribunal's jurisdiction was thus reduced to XOM's main claim regarding the expropriation of its 41⅔% interest in the Cerro Negro AA.

4.1.3.2 Non-compliance with the Treaty

However, XOM argued before ICSID that the BRV had illegally expropriated its investment and, consequently, had breached the BIT with the Netherlands. With this argument, XOM was not seeking compensation but reparation. It would then be entitled to be restored, economically speaking, to the position it should have been in had the alleged illegal act of the BRVV not occurred, and XOM would take full advantage of the very substantial and unexpected increase in oil prices that occurred after the date of the expropriation. In fact, this was XOM's main reason for insisting on the illegality of the act. In 2012, XOM estimated that the total amount could reach 16.6 billion US$ (ICSID XOM vs. BRV, 2014, §176).

But the Tribunal concluded that the BRV had indeed followed a proper formal legal procedure. Moreover, while it had not yet paid compensation, this was only because of the disagreement between the parties, and would definitely be resolved by the arbitration itself. In the words of the Tribunal:

> […] the mere fact that an investor has not received compensation does not in itself render an expropriation unlawful. An offer of compensation may have been made to the investor and, in such a case, the legality of the expropriation will depend on the terms of that offer. In order to decide whether an expropriation is lawful or not in the absence of payment of compensation, a tribunal must consider the facts of the case (ICSID: Award XOM v. BRV, 2014; §301).

Given the confidential nature of the negotiations, there was limited public information available. However:

> It was the Claimants' burden to prove their allegations concerning the position taken by Venezuela during the discussions regarding the compensation to be paid. It is not disputed that negotiations took place, and it has been established that Venezuela made proposals during those negotiations. It seems likely that there were discussions at the time on the method of valuation of the expropriated interests, on the relevance of the cap provisions referred to by Venezuela and on the exact amount of the compensation payable to the Claimants. The Tribunal finds that the evidence submitted does not demonstrate that the proposals made by Venezuela were incompatible with the requirement of "just" compensation [...] of the bilateral investment treaty. Accordingly, the Claimants have not established the unlawfulness of the expropriation on that ground (ICSID: Award XOM vs. BRV, 2014; §305).

Hence, the Tribunal rejected XOM's position about the nationalization being unlawful, and specified that the "compensation must be calculated in conformity with the requirements [...] of the BIT" (ICSID, Award XOM vs. BRV, 2014; §306).

4.1.3.3 Market Value and Threshold Price

Regarding the threshold price of 27 US$(1996)/b of Brent crude, the BRV argued that, in the best case scenario, the market value of the expropriated portion could be no other than that corresponding to the maximum level of profits guaranteed by PDVSA. Any profits beyond this level would have been fully exposed to the country's tax sovereignty. For its part, XOM insisted that the threshold price would only be relevant insofar as it limited PDVSA's obligations, whereas it was irrelevant as far as the international obligations of the BRV were concerned.

Following XOM's reasoning, the Tribunal decided to ignore the threshold price determined by the AA, and it quantified the compensation to be paid at 1.412 billion US$, based on a discount rate of 18%. In any case, according to the AA, there would be no question of double recovery, since XOM would have to return to PDVSA the amount it had already received based on the award of the International Chamber of Commerce, which had been set at 908 million US$. Thus, XOM would have obtained an additional indemnity of 504 million US$. However, the BRV immediately requested the annulment of the award.

4.1.4. Annulment (BRV vs. XOM)

ICSID awards are not subject to appeal, although a request for annulment can be made, which the BRV duly introduced in February 2015. In addition, the BRV requested that, in the meantime, enforcement of the award be suspended. ICSID proceeded to appoint an *ad hoc* Committee and the enforcement of the award was indeed suspended (ICSID BRV vs. XOM, 2017, §156).

The BRV moved for annulment on several grounds, the most important being the one related to the threshold price. Let us recall that in Condition Eighteen of the FOC of the Cerro Negro Project, the Venezuelan state reaffirmed its sovereign rights, and in Condition Twenty, the obligation was established in the Cerro Negro AA to set a threshold price with respect to PDVSA's commitment to indemnify the foreign investor. The Committee, in its Annulment Decision, quoted the Award, which stated that:

> The Dutch Treaty [...] includes "the law of the Contracting Party" and "the provisions of special agreements relating to the investments" among the sources of applicable law. In reliance on these provisions, the Respondent appears to argue that the Eighteenth and Twentieth Conditions of the Cerro Negro Framework of Conditions (i) relieve the Republic from its obligation to comply with the standards set forth in the Treaty, and/or (ii) import the contractual limitations to [PDVSA/Subs]'s indemnity obligations [...] into the State's responsibility for breach of the Treaty" (ICSID, Decision on Annulment BRV vs. XOM, 09/03/2017; §156. Emphasis ours).

'Appears', and nothing more. Thus, the Tribunal reasoned that the BRV's approach was unintelligible. Instead, as the Committee observed:

> [...] it should be noticed that Venezuela has vehemently denied in these Annulment proceedings that it ever did advance these propositions in its defence in the Arbitration, and says that the Tribunal is repeating here the miss-description of its position put forward in argument by the [Exxon]Mobil Parties (Quoted in ICSID, Decision on Annulment BRV vs. XOM, 2017; §179).

We continue with the Committee quoting the Tribunal's Award:

> The Tribunal disagrees with this position. The Tribunal recalls that it is a fundamental principle of international law that "[a] party may not invoke the provisions of its internal law as justification for its failure to perform a treaty". Under this principle, international obligations arising from a treaty cannot be discarded on the grounds of national law. [...] the Tribunal has no doubt in concluding that this issue must be governed by international law. Consequently, the Eighteenth and Twentieth Conditions cannot exempt or excuse the Respondent from its obligations under the Treaty or under customary international law. Bearing this in mind, the Tribunal has considered the effect of the Eighteenth and Twentieth Conditions of the Cerro Negro [FOC] in the section on quantum below (ICSID, Decision on Annulment BRV vs. XOM, 2017; §179).

Ultimately, the Tribunal in its Award had concluded that:

> The price cap contained [...] is applicable only to the compensation payable by [PDVSA/Subs]. Since the Respondent in this proceeding is the Bolivarian Republic of Venezuela, not [PDVSA/Subs], the Tribunal concludes that it may not oppose this price cap to the Claimants (ICSID, Decision on Annulment BRV vs. XOM, 2017; §143).

However, as the Annulment Committee observed, international law does not create or modify property rights. It can only create rights for its defense;

because as far as property rights are concerned, these are always defined by national law. In this sense, the Tribunal, first, completely omitted in its Award the conceptual determination of 'market value' – determined by a hypothetical interested buyer – and then established an inappropriate relationship between national and international law. Consequently, the *ad hoc* Committee annulled the award on this precise point but confirmed its broad general terms. That is to say, for the time being, the BRV would not be obliged to pay 1.412 billion US$; but the discount rate of 18%, for example, was confirmed. The *ad hoc* Committee specified:

> The Committee should not be understood by this Decision to be determining in what way either the Price Cap or the outcome of the International Chamber of Commerce arbitration should be brought to bear on the assessment of compensation in the present dispute. That would lie beyond its competence, and is in any event a decision which the Committee is in no position to make. The Committee's decision is directed simply at the a priori exclusion in the Award of certain essential elements from the process of arriving at the compensation due pursuant to the terms [...] of the bilateral investment treaty (ICSID, Decision on Annulment BRV vs. XOM, 2017 §190).

4.1.5. Going back to ICSID

According to the Committee's decision, XOM had the right to request ICSID arbitration again, albeit this time only to determine the impact of the threshold price on the compensation to be paid by the state. For an interested buyer, with an expectation of prices above the ceiling defined in the AA, the expected profits would essentially be divided into two tranches: the first one, based on the guarantee of PDVSA and Lagoven, and the second one, without the benefit of guarantee and therefore fully exposed to national tax sovereignty. How much would the proverbial 'interested buyer' be willing to pay for the latter tranche? XOM again initiated ICSID proceedings, and the new arbitral tribunal was constituted in August 2020.

4.2. ConocoPhillips

In the Petrozuata Association, ConocoPhillips (COP) had a 50.1% shareholding, and CVP 49.9%. In the Hamaca Association COP had 40%, and Chevron and CVP had 30% each. Consequently, in the case of the Hamaca AA, since COP refused to migrate, Chevron was to keep its 30% in the new Joint Venture; however, since COP was the only partner in Petrozuata, PDVSA was to assume full ownership of the project. The state, for its part, had to compensate COP for the expropriation of its shares in the two associations. At the outset, COP did not resort to arbitration before the ICC against PDVSA, and only initiated ICSID proceedings against the BRV.

4.2.1. International Center for Settlement of Investment Disputes (COP vs. BRV)

The Venezuelan government and COP started negotiations on the case already before June 26, 2007. Nevertheless, COP was going through the formalities of initiating arbitration against the BRV before ICSID. The two parties appointed their respective arbitrators and, in the absence of agreement between them, ICSID appointed the president of the tribunal. After the parties submitted their respective memorials, the first hearings took place between May and July 2010. On the basis of these hearings, in September 2013, the Tribunal rendered a Decision on Jurisdiction and the Merits (ICSID COP vs. BRC, 2013).

4.2.1.1. Decision on Jurisdiction and Merits

In the first part of its Decision on Jurisdiction, the Tribunal followed essentially, and unanimously, the same furrows already plowed in XOM vs. BRV. However, as to the second part on the *Merits*, the Decision was taken by a majority consisting of Kenneth Keith, the presiding arbitrator, a New Zealand national appointed by the World Bank, and Yves Fortier, an arbitrator appointed by COP, a Canadian national. The BRV-appointed arbitrator, Georges Abi-Saab, a national of Egypt, penned a *Dissenting Decision*.

The expropriation took place on June 26, 2007. However, just after that, oil prices on the world market unexpectedly began to rise sharply, to levels seemingly destined to endure for a long term. Thus, the fair market price, based on July 2007 market expectations, was much lower than the fair market price with subsequent expectations of much higher prices. If the expropriation were to have been deemed unlawful, then COP would not be entitled to *compensation*, but to *reparation* to be quantified at the date of the award. Of course, the reparation would have been such as to nullify all the consequences of the unlawful act and re-establish the situation that would have existed had the act not occurred (ICSID COP vs. BRV, 03/09/2013, §338). Consequently, COP would fully benefit from the high prices subsequent to nationalization. The Majority ruled in favor of the date of the award as the valuation date (ICSID COP vs. BRV, 2013, §212).

However, the Majority also noted that, in order for an expropriation to be lawful, an immediate payment at the precise moment of expropriation was not necessary:

> The requirements for prompt payment and for interest recognise, in accordance with the general understanding of such standard provisions, that payment is not required at the precise moment of expropriation (ICSID, Decision on Jurisdiction and the Merits COP vs. BRV, 2013; §362).

In fact, at a meeting held in March 2007 between COP and Venezuelan representatives, Venezuela offered compensation of 1.1 billion US$ for its participation in Petrozuata, and 1.2 billion US$ in the case of Hamaca, in the event that COP did not migrate. That is to say, the BRV offered in total 2.3 billion US$. COP's representatives considered that this offer was "far below the fair market value of our interests in the Projects" (ICSID, Decision on Jurisdiction and the Merits COP vs. BRV, 2013; §389).

In any event, the expropriation would only have been unlawful had Venezuela's proposal been so inadequate as to constitute a *de facto* refusal to pay the compensation due under the BIT. However, the Majority would not analyze the quantum of compensation from this perspective. Instead, immediately following the lines just quoted, the award stated:

> But it is also commonly accepted that the Parties must engage in good faith negotiations to fix the compensation in terms of the standard set, in this case, in the BIT, if a payment satisfactory to the investor is not proposed at the outset (ICSID, Decision on Jurisdiction and the Merits COP vs. BRV, 2013; §362).

Then, as evidence of the lack of good faith on the part of the BRV, the Decision referred to a COP Brief dated September 15, 2008:

> In their Memorial the Claimants state that the 29 March verbal offer represented no more than 5 % of the real value of their investments and, therefore was totally inadequate (ICSID, Decision on Jurisdiction and the Merits COP vs. BRV, 2013; §390).

A comparison was drawn between an offer originally made in March 2007, and what COP considered as the 'real value of their investments' in its September 2008 Memorial. This 'real value', based on the assumption of an unlawful expropriation, COP was to estimate "provisionally" at 30.3 billion US$. The Majority thus accepted COP's position that this was a 'totally inadequate' offer by the BRV.

The Majority then concluded that:

> The Respondent breached its obligation to negotiate in good faith for compensation for its taking of the ConocoPhillips assets [...] on the basis of market value as required by [...] the BIT;

and, consequently:

> The date of valuation of the ConocoPhillips assets is the date of the Award (ICSID, *Decision on Jurisdiction and the Merits* COP vs. BRV, 2013; §404).

In short, the BRV's actions had been shown to be lacking in the good faith which, according to the Majority, would have required it to accommodate COP's position on compensation in the negotiations.

4.2.1.2. Dissent

Abi-Saab in his *Dissenting Decision* (ICSID COP vs. BRV, 19/02/2015, *Dissenting Decision*), first of all, stressed that even the initial offer made by

BRV, of 2.3 billion US$, was very reasonable. Indeed, this figure corresponded to the evaluations that, in 2006, had been made by ConocoPhillips itself regarding the market value of its stakes in Petrozuata and Hamaca. These valuations were, in fact, among the documents in the record submitted by ConocoPhillips (ICSID COP vs. BRV, 19/02/2015, *Dissenting Decision*, §156 and §159).

But there was something else. It was not the case that it was 'commonly accepted that the Parties must engage in negotiations in good faith'. The legal basis for the arbitration in question was the Venezuela-Netherlands BIT, which contained no reference to an obligation to negotiate compensation in 'good faith', indeed, the treaty did not even refer to the obligation to negotiate. Nevertheless, the BRV had made a reasonable first offer and left no doubt as to its willingness to submit to ICSID arbitration proceedings, in accordance with the Venezuela-Netherlands BIT, in order to reach a final result.

Moreover, the Majority, with its decision regarding the lack of good faith in the negotiations on the part of the BRV, had attributed to itself a power that it did not have. Indeed, none of the parties involved in the arbitration had ever made an accusation of bad faith against the BRV. Consequently, it was *ultra petita*. The Majority had exceeded its competence, in a manner that did not even give the BRV the opportunity to defend itself; thus the Majority had violated the BRV's procedural rights (ICSID COP vs. BRV, 19/02/2015, *Dissenting Decision*, §284f).

4.2.2. WikiLeaks

As soon as the Award in question was published, the BRV requested that the Tribunal reconsider the majority decision for legal reasons specific to it, but also because an extraordinary and unexpected event had occurred in the meantime. WikiLeaks published on the Internet a cache of diplomatic communications from all U.S. embassies worldwide. In particular, there were the reports from the U.S. Embassy in Caracas, which reported on the meetings held on several occasions in 2008 between Embassy officials and COP representatives, and which made reference to the course of negotiations. In one of these cables, dated April 4, 2008, the U.S. Ambassador reported a conversation with Greg Goff, COP's chief negotiator:

> Goff stated the BRV [...] has agreed on a fair market methodology [...] for computing the compensation for the expropriated assets. However, given the recent increase in oil prices, the fair market value of the assets has increased. [...] COP has proposed a settlement number and the BRV appears to be open to it. Goff added that COP also plans on increasing the settlement number [...] due to recent increases in oil prices.

In fact, the negotiations were bound to fail, because every time Venezuela wanted to close the gap, COP backed away and increased its demands in line with the rise in oil prices.

Goff, a witness at the quantum hearings in August 2016, fully confirmed this summary (BRV, ICSID, Application for Annulment BRV vs. COP, 08/07/2019; §54). However, the Majority did not change its Decision concerning the lack of good faith on the part of the BRV in the negotiations.

4.2.3. Abandoning Appearances

By the time the Majority was finalizing its Decision, Abi-Saab was already seriously ill, forcing him to spend months in hospital, throughout 2014. However, on September 3, 2013, he had agreed with the Majority that their Decision could be issued without his signature and that he would deliver his Dissenting Opinion at a later date. In fact, he was only able to finish it on February 19, 2015.

He also agreed with the Majority that, upon completion of his Dissenting Opinion, he would resign immediately because his health did not allow him to continue in the case. However, when Abi-Saab eventually submitted his resignation on February 20, 2015, surprisingly the members of the majority rejected it. In view of such rejection, and in accordance with the ICSID Convention, the president of the World Bank would have to designate a new arbitrator, not the BRV. In August, the World Bank appointed Andreas Bucher, a Swiss national, as the third member of the tribunal. Now, the BRV, the respondent, had no longer its own arbitrator.

4.2.4. Award and Request for Annulment

In March 2016, Kenneth Keith, the president of the tribunal, surprisingly resigned without explanation. In April, the World Bank appointed Eduardo Zuleta, a Colombian national, as his replacement. Thus, out of the three members of the tribunal in 2013, only Yves Fortier, the COP-appointed arbitrator, remained on the tribunal by the end of 2015, with both Bucher and Zuleta having been subsequently appointed by the World Bank president. This put Fortier in an extraordinarily powerful position.

The Tribunal issued its final award in February 2019 (ICSID COP vs. BRV, 2019). The amount of reparations was calculated based on a discount rate of 17.25%. The Petrozuata reparations were set at 3.371 billion US$, and the Hamaca reparations were set at 4.442 billion US$, for a total of 7.813 billion US$.

In July 2019, the BRV requested the annulment of the Award, partly on the same grounds as in the XOM case – ignoring the threshold prices – but also on grounds of *ultra petita*, as to the alleged lack of good faith on the part of the BRV in the negotiations – and, therefore, the illegality of the expropriation –, and on the fact of having been stripped, in 2015, of the right to appoint an arbitrator. Of course, the BRV also requested a stay of enforcement of the Award.

4.2.5. *International Chamber of Commerce (COP vs. PDVSA)*

In 2014, COP finally resorted to the International Chamber of Commerce to claim compensation from PDVSA under the AA. In its Award, in 2018, the ICC tribunal applied a discount rate of 18%, and thus established an indemnity of 489 million US$ for Petrozuata, and 1.466 billion US$ for Hamaca, for a total of 1.955 billion US$, effective for May 2016. Thereafter, PDVSA agreed with COP on the procedure for the corresponding payment.

4.3. Conclusions

The arbitrations filed by XOM and COP before the International Chamber of Commerce against PDVSA, based on the respective AAs for the production and upgrading of extra-heavy crude oil from the Orinoco Oil Belt, yielded sensible results for all parties. In the XOM/Cerro Negro case, the corresponding arbitration tribunal quantified, in December 2011, the compensation due at 908 million US$. In the COP/Hamaca case, in May 2016, the corresponding tribunal quantified it at 1.466 billion US$. As the threshold prices were practically the same, this difference is essentially explained by the fact that Hamaca upgraded the crude to higher API degrees, and also processed higher volumes. On the other hand, the compensation in the COP/Petrozuata case was comparatively small at only 489 million US$, due to the low threshold prices in this first AA.

However, with respect to the arbitrations filed by XOM and COP against the Bolivarian Republic of Venezuela before ICSID, based on the Venezuela-Netherlands BIT, the situation was very different. In both cases, the tribunals, in line with submissions made by claimants, decided to disregard the threshold prices defined in the AAs, when quantifying their fair market value. Hence, the tribunal, in the XOM/Cerro Negro case, quantified the compensation due, in 2014, at 1.443 billion US$. The BRV successfully resorted to an annulment proceeding, on the grounds that the tribunal had not taken into consideration the threshold prices, without giving any reasons. In 2020, XOM again filed another arbitration against the BRV, but now strictly limited to the issue of threshold prices. This arbitration culminated in July 2023 with an award that essentially vindicated the position that the BRV had taken since 2007[2].

In addition, XOM and COP requested the respective tribunals to declare that the expropriation had been unlawful. However, the tribunal in the XOM case ruled that it had been lawful, but the COP tribunal ruled that it had been unlawful. Consequently, the valuation date in the COP case would be the date

---

[2] In its decision of July 10, 2023, the Tribunal concluded that the threshold prices did apply. Thus, the 504 million US$ struck off by the Annulment Decision were reduced by 419 million US$, leaving the BRV with a balance of only 85 million US$ to pay for this concept.

of the corresponding award, i.e. in 2018, with the claimant thereby benefiting from the unexpectedly high prices after July 2007.

We note that in the case of XOMM, the compensation to be paid by PDVSA pursuant to the ICC proceedings was 908 million US$, whereas the compensation to be paid by the BRV pursuant to the ICSID proceedings came to 1.443 billion US$, an increase of 59%. In the case of COP, the respective figures are 1.955 billion US$ and 7.813 billion US$, i.e., an increase of 300%. This shocking percentage difference essentially reflects the fact that nationalization was declared unlawful in the case of COP, which was not the case for XOM.

Also in the COP case, the BRV resorted to annulment proceedings, not only because of the threshold prices, but also because the expropriation was declared unlawful and the BRV was deprived of its right to appoint its own arbitrator. The annulment proceedings are still ongoing.

5. From the European Energy Charter to the Energy Charter Treaty

The fall of the Berlin Wall in November 1989 marked the starting point of a process that culminated in the dissolution of the Soviet Union in December 1991. This was the greatest political crisis of the late 20th century, which, apart from coming as a complete surprise, was also a major economic crisis for the entire area, both in what is now the Russian Federation, as well as in the countries of Central Asia and Eastern Europe which, up until then, had been part of the USSR.

However, as early as June 1990, the Prime Minister of the Netherlands, Ruud Lubbers, proposed at the European Council to create an institutional framework based on complementary relations in energy policy between the two areas: on the one hand, the Eastern Bloc and, on the other, Western Europe. The abundant natural energy resources of the former could mobilize the abundant financial and technological resources of the latter, and thereby become the axis of a new relationship that would contribute to overcoming the deep crisis affecting the Eastern Bloc, and thereby facilitate the transition of its economies from communism to capitalism.

At that time, the Soviet Union was the second largest petroleum exporter in the world, although it only held 6% of proven reserves, whereas OPEC held 76%. However, in terms of natural gas reserves, the situation was very different. Only 42% of these were in OPEC, followed closely by the Soviet Union with 39%.

In February 1991, the two groups of countries began negotiations. The USA, Canada, Australia and Japan soon joined in. The negotiations ended in December of the same year with a political declaration under the name of the *European Energy Charter*, which was signed by all participants. Negotiations on what was to become the Energy Charter Treaty (ECT) were then initiated and concluded in December 1994. The corresponding *Secretariat* was created in Brussels.

The ECT would enter into force, provisionally, in all signatory countries even before it was approved and ratified by the respective countries. In particular, the clauses relating to international arbitration came into force immediately. According to its promoters:

> These provisions will reassure foreign investors as to the fair, consistent and predictable interpretation and application of the ECT rules by all Contracting Parties.
>
> If the investor of another Contracting Party considers that a Government has not fulfilled its obligations prescribed by the provisions of the Investment Protection, the investor may, with the unconditional consent of the Contracting Party, [...] submit it to international arbitration. [...] The Contracting Parties have undertaken [...] to implement arbitration decisions without delay and to ensure that they will be effectively enforced in all their territories (ECT Secretariat, 1995, p. 11).

The Treaty was to cover all activities and investments in the energy sector, from exploration, extraction, refining, production and storage, including land and maritime transportation, through to the distribution and marketing of energy products.

The conception of the Treaty essentially corresponds to the interests of the OECD member countries, as petroleum-importing countries and with the British reference in mind. However, the OECD does include one oil-producing country: Norway. It was on account of the presence of this country in the negotiations that in Article XVIII, under the title of *Sovereignty over Energy Resources*, the sovereign rights of nation states were reaffirmed:

> Each state continues to hold [...] the rights to decide the geographical areas within its Area to be made available for exploration and development of its energy resources, the optimization of their recovery and the rate at which they may be depleted or otherwise exploited, to specify and enjoy any taxes, royalties or other financial payments payable by virtue of such exploration and exploitation, and to regulate the environmental and safety aspects of such exploration, development and reclamation within its Area, and to participate in such exploration and exploitation, inter alia, through direct participation by the government or through state enterprises (Art. 18.3).

However, while Norway – like energy-rich Australia – signed the Energy Charter Treaty in 1994, they would never go on to ratify it. For their part, the USA and Canada did not even deign to sign: rich in energy resources, with a dominant position in the Western Hemisphere, they were simply unwilling to submit the management of their resources to international arbitration.

5.1. Russian Federation and G8 Ministerial, Moscow 1998

In April 1998, the Ministerial Conference of the 'G8', which grouped together the seven most important capitalist economies in the world, plus the Russian Federation, was held in Moscow. The ECT had just reached the number of approvals and ratifications required for its entry into force. Among the approvals were those of all the member countries of the European Union, as well as that of the European Union itself. However, the Russian Federation had not approved it, and the *ECT Secretariat* and the *International Energy Agency* now submitted a joint paper, under the title *Energy Investment*, to convince the Russian Federation to approve and ratify it (ECT Secretariat and IEA, 01/04/1998).

This paper expressed the conviction that, in a country with an economy in transition, where the appropriate legal and fiscal regimes were yet to be developed, there would be no doubt that investor confidence would require, first and foremost, the approval of the ECT with its international arbitration clauses. Moreover, while in the past, in many countries, the scope for private investment in the energy sector had been limited as a result of the involvement of governments and state-owned enterprises, the situation had changed:

> There is now widespread recognition that liberalisation, including privatisations and reduction of monopolies yields major benefits for the efficient operation of energy markets (ECT Secretariat and IEA, 01/04/1998; p. i).

The paper also reported on foreign investors' criticisms of the Russian Federation and claimed that more comprehensive studies would show that the Russian petroleum regime was inadequate. Indeed, Joint Ventures in the country were based on an administrative system of 'Subsoil Licenses'. This document was considered to be of paramount importance, while the arrangements between the parties were of only subordinate importance:

> This exposes the investor to several significant risks. The terms of the license to use the subsoil are subject to unilateral change by new legislation and are terminable by the government on various grounds. It is subject to all applicable taxes at all levels of government, and no protection is provided against adverse changes in tax laws or other laws. There is no clear right to export, and disputes are not subject to impartial adjudication because there is no contractual relationship between the Joint Venture partners and the government (ECT Secretariat and IEA, 01/04/1998, p. 20).

The Russian government granted concessions safeguarding its sovereign rights in petroleum matters, on which the joint ventures were based. It was suggested that a different procedure ought to be used instead: granting the state's right of access to a state-owned company and then having the latter make binding commitments to the foreign private partner:

> Risk of fiscal change is a major concern to investors making large upfront investments, especially in regions with little or no history of fiscal stability. Frequent tax changes are due in part to the nature of gross-revenue based regimes where governments need to make adjustments to benefit from changes in prices or cost. Profit-based systems are more self-adjusting and give a better basis for investors to assess the fiscal impact over the life of their investment project. Such systems do not impose a heavy tax burden in the early years of production which would negatively affect projected rate of return (ECT Secretariat and IEA, 01/04/1998, p. 21f).

Clearly, the goal was to eliminate the royalty, and replace it with a tax on excessive profits. Evidently, the petroleum regime proposed by the Energy Charter Secretariat and the IEA was the same as the one already being implemented in Venezuela.

It should be noted that OPEC member countries had not been allowed to participate in the negotiations that were to culminate in the ECT. Historically, they had created a political and legal structure that had to be dismantled before they could be forced to become members of the ECT, once the latter was properly structured and consolidated. Nevertheless, a representative of PDVSA was allowed to participate as an observer.

5.2. Yukos

Between 1996 and 2003, the Russian Federation initiated a process of privatization of part of its state-owned petroleum industry, and the company Yukos emerged therefrom as the country's largest oil and natural gas producer. However, in 2003, a number of major shareholders and senior executives of Yukos were arrested on charges of massive tax fraud, falsification of documents and other similar offenses. As a result, between 2004 and 2007, the state seized most of Yukos and transferred it to a state-owned petroleum company, Rosneft.

Starting in 2005, former Yukos shareholders initiated an international arbitration case based on the ECT, per the rules established by the United Nations Commission on International Trade Law. The Tribunal was established in 2007, sitting in The Hague; and in November 2009 it unanimously took an *Interim Decision on Jurisdiction and Admissibility*. Since the Russian Federation had not ratified the Treaty, it only applied provisionally:

> […] to the extent that such provisional application is not inconsistent with its constitution, laws or regulations (ECT, Art. 45).

In line with the doctrine of Kompetenz Kompetenz, it was for the Tribunal itself to decide its own jurisdiction and, in its Interim Decision, it concluded that there would be no contradiction between the arbitration clauses of the ECT and the constitution, laws or regulations of the Russian Federation. Already in

August 2009, the Russian Federation had officially and formally informed the *Energy Charter Secretariat* that it would definitely not approve the ECT. Nevertheless, according to the same Article 45 of the ECT, the arbitration clauses were to remain in force for another twenty years, in respect of investments made prior to this date, i.e. until August 2029, and this was stated in the Decision in question.

The issue of tax evasion was not on the ECT's list of issues that might being subject to arbitration. But the claimants argued that this was only a pretext for the Russian government to expropriate their assets. Thus, the issue in dispute was an unlawful nationalization and this would be subject in the ECT to conditions identical to that have already been explained in relation to the Venezuela-Netherlands Treaty. The Tribunal accepted the position of the claimants and concluded that the Russian tax measure were, in fact, subject to arbitration.

There was still the fact that Yukos had originally been a Russian company and, as such, its owners would not have had access to international arbitration under a treaty. However, prior to the arbitration, the shareholders had reorganized their corporate structures, which meant that the claimants were two companies, one from Cyprus, one from Great Britain, both countries that had ratified the ECT. The Tribunal concluded that these were foreign investors, and that the case was admissible.

According to the claimants, the actions of the Russian state amounted to an illegal expropriation and, consequently, they were entitled to reparations estimated at 100 billion US$. The tribunal, in its July 2014 Award, arrived at a figure of 50 billion US$, a huge amount, since, in the past, no international arbitration had reached even 4% of that figure.

The Russian Federation filed for the annulment of the award, in accordance with the law of the Netherlands. The Dutch court ruled in favor of Russia, but the claimants appealed to the next instance, which ruled against Russia. Russia then appealed to the Supreme Court, which ruled, in November 2021, that the award would remain suspended, and had to be reconsidered.

5.2.1. Yukos and Venezuela

The law firm that took on the case for Yukos was *Shearman and Sterling*, while representing the Russian Federation was *Cleary, Gottlieb et al.* These were the two key law firms, each of which, in the 1990s, had assisted PDVSA with the Oil Opening policy. Appointing *Shearman and Sterling* on behalf of Yukos made sense because that firm specialized in defending the interests of investors. But the fact that Russia appointed *Cleary, Gottlieb et al.* which also specialized in defending investors, can only be explained by the fact that the Russian government lacked any experience in the field of international arbitration.

Yukos appointed as its arbitrator Charles Poncet, a Swiss national, and Russia appointed Stephen Schwebel, a United States national (Schwebel had assisted both COP and XOM as an outside legal consultant, sitting as an arbitrator in a moot court, to prepare the legal teams of both companies in anticipation of their respective proceedings against the BRV before ICSID. The Permanent Court of Arbitration appointed Yves Fortier as president of the Tribunal, a Canadian national. Fortier was also the arbitrator appointed by COP before ICSID in Washington in the COP vs. BRV case.

5.4. Vattenfall and Germany

The *European Energy Charter* originally arose from the dissolution of the Soviet Union, so it might be thought that, with the Russian Federation's decision not to participate in the ECT, this treaty amounted to a failure. But the ECT eventually morphed into a multilateral energy investment treaty that, in any case, would serve foreign investors in the energy field to keep in check the aspirations of nation states to an international ground rent as owners of the corresponding natural resources.

Furthermore, such a treaty would guarantee rights to foreign investors beyond the rights of domestic investors. The latter would not have access to international arbitration.

For example, following the disaster at the Fukushima nuclear power plant in March 2011, the German Federal Parliament decided in June 2011 to phase out nuclear power within the next eleven years, but to immediately shutting down the two oldest plants, already scheduled for decommissioning because they were considered too risky to allow them to continue operating.

This decision pitted, on the one hand, a structure of economic interests in which the state was involved and, on the other hand, the sovereign state, responsible as such for environmental regulation. Therefore, it was necessary to delimit the economic rights of companies and separate them from the sovereign rights of the state – and even from its obligations as Sovereign – in terms of environmental regulation. Of course, the generating companies involved would be entitled to compensation for the impairment of their assets.

However, the two power plants in question were owned by Vattenfall, a Swedish company that, on the basis of the ECT, in 2012 initiated an ICSID arbitration. At ICSID, entirely in accordance with the ECT, the state would be considered the sole guarantor of the foreign investor which, in the case of measures amounting to an expropriation, would be entitled to receive compensation equal to the market value of these two plants before the measure in question became known, i.e. before the Fukushima disaster. This would entitle Vattenfall not only to compensation for the asset impairment, but also for lost profits. Definitively, Vattenfall sued the Federal Republic of Germany for 4.7 billion US$.

The consequence of this lawsuit would be that the German people and the German business community were subject to the development of environmental policies, qua sovereign policies in themselves, and would share in bearing their costs, whereas foreign companies would not. The latter, thanks to the ECT, would simply not be subject to the sovereign rights of the contracting states with regard to their protected investments in the energy sector.

However, the European Community, due to the progressive integration of its members into a common legal framework, had for years been questioning the validity of BITs between member countries. In 2020, 23 member countries agreed to put an end to the BITs currently in force between them. However, the ECT was an international treaty, with a scope extending beyond the European Union. Finally, in 2021, the EU Commission decided that the ECT would also not apply to disputes between investors and governments of EU member countries. However, in the Vattenfall case, the German government settled with the investor, and agreed to a compensation of 1.5 billion US$, much more than what Vattenfall would have been entitled to receive under applicable German law.

6. Multilateral and Bilateral Investment Treaties of the OECD

The ECT entered into force in April 1998, after having been approved and ratified by thirty countries. Its promoters then began negotiations for a *Multilateral Agreement on Investment* (MAI) among OECD member countries, dealing with investment in general. That same month, the G8 Ministerial Conference took place in Moscow, and in their joint paper, *Energy Investment*, the ECT Secretariat and the IEA insisted that the Russian Federation should also become a member of the MAI, once the OECD negotiations were completed.

However, the success of the ECT negotiations, in so far as the OECD was concerned, came down to the management of the political dimensions of the discussion. The ECT and its international arbitration clauses, as a practical matter, were only supposed to apply to disputes between underdeveloped and former communist countries, and not to disputes between OECD member countries, as the latter were considered implausible.

The proposed MAI would have covered investments in general, within the framework of the OECD member countries, as opposed to only investments in the field of energy. Third World countries, as well as former communist countries, would subsequently have been allowed to join the Agreement, but they would have been required to accept the Agreement as it stood, with the latter having been negotiated solely by the developed capitalist countries. By December 1998 there was a complete draft of the MAI, with international arbitration clauses incorporated (essentially the same as those already incorporated in the ECT). However, France put an end to the negotiations with a resounding *no* to the draft, as it considered that such clauses were incompatible with its status as a sovereign country.

6.1. Canada-EU Comprehensive Economic and Trade Agreement: Dispute Settlement

In 2004, the European Union began negotiating a BIT with Canada, the *Comprehensive Economic and Trade Agreement* (CETA). Negotiations were completed with a proposed agreement in 2014. After passing through the European Parliament and the European Court of Justice, the agreement entered into force provisionally in September 2017. For it to become definitively effective, it had been approved and ratified by all EU member countries, a process that was predictably going to take years. To date, 15 of the 27 EU countries have already approved and ratified it.

However, the provisional validity of CETA *excluded* the settlement of disputes between foreign investors and the respective nation states. However, once the agreement was duly approved and ratified by all EU member countries, a variety of forums and rules were contemplated for the settlement of disputes by international arbitration under the agreement: ICSID, the Stockholm Chamber of Commerce, or *ad hoc* proceedings in accordance with the rules of commercial arbitration approved by the United Nations, or even some special arrangement agreed between disputing parties in a particular case. In what follows, we will focus on ICSID.

CETA was to create a Joint Committee, which would have dealt with all controversies that might arise between the Parties. First, upon entry into force of the Agreement, this Committee was to appoint a tribunal of fifteen duly qualified members: five EU nationals, five Canadians and five third-country nationals. The respective arbitrators were to be chosen at random, if necessary. The parties would no longer have the right to appoint a specific arbitrator; they would now have to settle for an arbitrator chosen from the members of the panel approved by the Joint Committee. These arbitrators were not allowed to act as counsels in any arbitral proceedings.

Moreover, a second instance was created, so that the final award of the first instance could be submitted to an appellate tribunal, which, in turn, would be appointed by the Joint Committee, also using a random selection procedure. This, without prejudice to the possibility that a party might request the annulment of an award under the ICSID procedure (CETA, Sections D and E).

Ultimately, in the formation of the ICSID arbitration tribunals, the Joint Committee adopted rules meant to address the most notable abuses that Tribunals had perpetrated in cases involving Third World and former communist countries, albeit essentially without interfering with the ICSID mechanism itself. In short, the idea was to keep ICSID fully in force vis-à-vis these groups of countries.

7. Globalization and Natural Resources: Conclusions

In the course of the 20th century, the finiteness of the globe became increasingly relevant, leading to a confrontation over extraordinary surpluses between raw material exporting countries and importing countries. This finiteness also led to an increase in production costs due to the massive destruction of nature and thus of the habitat.

It has now become unavoidable to implement policies to halt this destruction. However, arresting the destruction of nature and the natural habitat, free gifts of nature as such, implies significant costs, but not doing it will be even more costly. For the time being, the response of capitalism, in its process of globalization, consists of denying the sovereign rights of states – of developed and underdeveloped countries alike – and, as the driving force of this process, international capital has been claiming the right to use nature, including by destroying it, as a vested right. Through investment treaties, states are simply turned into one of two parties to a business relationship, which consists of attracting foreign investment in exchange for guaranteeing them certain conditions, regardless of any consideration of their rights and obligations as Sovereigns.

The development of international law based on international arbitration is a profoundly undemocratic process. With the development of democracy in the 20th century, the presence of states as sovereigns in the respective national Supreme Courts became increasingly important, and in the process the specific rights of individuals were delimited. In contrast, in the process of globalization, the concept of national sovereignty disappears but it is not superseded by the concept of an *international* and, therefore, *global* sovereignty, which would make it possible to delimit private rights, on the one hand, from sovereign rights, on the other.

Throughout the world, national policies continue to be considered – although to varying degrees – part of the traditional confrontation between labor and capital, between Left and Right, despite the growing importance and necessity of a policy for the defense of nature and the environment, which necessarily transcends the national sphere. It is therefore necessary that both left and right-wing policies incorporate increasingly important components of *international* and therefore *global* politics.

Failure to face this inescapable challenge has meant the catastrophic destruction of the political regimes – and, therefore, of the national economies – of many underdeveloped countries, giving rise to desperate emigration, which is repressed, sometimes brutally, by the developed countries. Moreover, there is also a growing frustration in the political processes of the developed countries, as it becomes increasingly obvious that their national governments are no longer in a position to solve the most urgent problems of their respective country and people, which are inherently *international and global* in character.

A properly *national* policy must aim at the elimination of the international arbitration structure and, together with this, at the revision of the structure associated with investment treaties. In their essence, these structures threaten the sovereign rights over natural resources and the environment of the peoples that inhabit the planet. International and global capital must be placed in its rightful place and made subject to the sovereign rights of the parties involved and, therefore, to the global sovereignty that will emerge from the corresponding international agreements.

Only in this way can the current generations ensure that they will bequeath the globe to future generations in an improved condition. The history of the development of capitalism and its relationship with natural resources shows that this policy will require many compromises, and will have to travel long and tortuous paths at national, regional and global levels. For the time being, however, the process of globalization underway is being led by international capital, which is assuming extreme, intransigent and definitely unsustainable positions.

Bibliography

ACEDO PAYAREZ Germán, *Jurisprudencia Petrolera Venezolana 1918-1971*, Caracas, Editorial Sucre, 1972.

ACOSTA HERMOSO Eduardo H., *Análisis Histórico de la OPEP*, vol. 1, Mérida (Venezuela), University of the Andes, 1969.

ADELMAN Morris A., "Efficiency of Resource Use in Crude Petroleum", *The Southern Economic Journal*, vol. 31, N° 2, 1964.

ADELMAN Morris A., *The World Petroleum Market*, Baltimore, Johns Hopkins University Press, 1972.

ADELMAN Morris A., "Politics, Economics and World Oil", *The American Economic Review*, vol. LXIV, 1974.

ADRIANI Alberto, *Labor Venezolanista*, Caracas, 1937 [1ère ed.].

ADRIANI Alberto, *Labor Venezolanista*, Caracas, Tipografía Garrido, 1937 [1st ed., revised 2nd ed. 1946].

AKINS James E., "The Oil Crisis : This Time the Wolf is Here", *Foreign Affairs*, vol. 51, N° 3, 04/1973.

ALASKA, *Statutes, Oil and Gas Leasing*, §38.05.180, 1995.

AL-KASIM Farouk, *Managing Petroleum Resources : The 'Norwegian Model' in a Broad Perspective*, Oxford, Oxford University Press, 2006.

ARMAS Celestino, Minister of Energy and Mines, *Comunicación al Presidente de PDVSA, Andrés Sosa Pietri*, 21/09/1990.

ARMAS Celestino, Minister of Energy and Mines, *Comunicación al Presidente del Congreso de la República,* 14/06/1991.

ARRIOJA José Enrique, *Clientes Negros. Petróleos de Venezuela bajo la Generación Shell*, Caracas, 1998.

BAPTISTA Asdrúbal et MOMMER Bernard, "La Economía Política de Venezuela", in Asdrúbal BAPTISTA, *Itinerario por la Economía Política*, Caracas, Ediciones IESA, 2008.

BAPTISTA Asdrúbal, "Más allá del Optimismo y del Pesimismo : las Transformaciones Fundamentales del País", in NAÍM Moisés et PIÑANGO Ramón (eds.), *El Caso Venezuela - una Ilusión de Armonía*, Caracas, Ediciones IESA, 1984.

BAPTISTA Asdrúbal et MOMMER Bernard, *El Ingreso Petrolero y la Capacidad de Absorción de Capital : el Caso de Venezuela*, Caracas, IESA, monograph, 1985.

BAPTISTA Asdrúbal et MOMMER Bernard, "El Petróleo en las Cuentas Nacionales : Una Proposición", *Revista del Banco Central de Venezuela*, april-june, 1986.

BAPTISTA Asdrúbal et MOMMER Bernard, *El petróleo en el pensamiento económico venezolano. Un ensayo.* Prologue Arturo Uslar Pietri, Caracas, Ediciones IESA, 1987.

BAPTISTA Asdrúbal (e.), *Venezuela Siglo XX : Versiones y Testimonios*, 3 vol., Caracas, Fundación Polar, 2000.

BAPTISTA Asdrúbal, *Bases Cuantitativas de la Economía Venezolana : 1830-2008*, Caracas, Fondation Polar, 2011.

BETANCOURT Rómulo, *Venezuela, Política y Petróleo*, Mexico, Fondo de Cultura Económica, 1956.

BETANCOURT Rómulo, *Posición y doctrina*, Caracas, 1959.

BETANCOURT Rómulo, *Tres años de gobierno democrático 1959-1962*, vol. III, Caracas, Editorial Cordillera, 1962.

BOUÉ Juan Carlos, *Venezuela – The Political Economy of Oil*, Oxford, Oxford University Press, 1993.

BOUÉ Juan Carlos, *The Political Control of State Oil Companies. A Case Study of the Vertical Integration Programme of Petróleos de Venezuela (1982–1995)*, doctoral thesis, Oxford, 1998.

BOUÉ Juan Carlos, *El Programa de Internacionalización de Petróleos de Venezuela, S.A.*, Caracas, 2004.

BOUÉ Juan Carlos, *A Question of Rigs, of Rules, ¿or of Rigging the Rules ? Upstream Profits and Taxes in US Gulf Offshore Oil and Gas*, with the assistance of Edgar Jones, Oxford, Oxford University Press, 2007.

BOUÉ Juan Carlos, *El Síndrome de la Orimulsión*, Caracas, Le Monde Diplomatique, 2012.

BOUÉ Juan Carlos, "Reflections on the Revolution in Britain (and on the Proceedings in certain Countries relative to that Event) : The Governance of Upstream Petroleum in the UK North Sea", to be published, 2019.

BOUÉ Juan Carlos, "Much More than a Footnote (or Three) : Frank C. Hendryx and an Untold Story of Petroleum Concessions and the Genesis of ICSID", *Journal of International Dispute Settlement*, vol. 13, Oxford, 2021.

BURDETT A.L.P. (ed.), *OPEC Origins and Strategy, 1947-1973*. Vol. 1: *1947-1959. Developments and Events Leading to the Creation of OPEC in 1960*, Cambridge, Archive Editions, 2004.

Canada-European Union Comprehensive Economic and Trade Agreement (CETA) 2017.

Nacionalización del Petróleo en Venezuela. Tesis y Documentos Fundamentales, Caracas, Catalá / Centauro Editores, 1975.

CHÁVEZ Hugo, *Una Revolución Democrática*, Caracas, 1998.

COMISIÓN BICAMERAL DE ENERGÍA Y MINAS, VENEZUELA, *Informe Cristóbal Colón*, 07/07/1993.

COMISIÓN BICAMERAL DE ENERGÍA Y MINAS, VENEZUELA *Informe del Estudio de las Asociaciones Estratégicas de PDVSA, sobre los Proyectos MARAVEN-CONOCO, y*

MARAVEN-TOTAL-ITOCHU-MARUBENI para la explotación y mejoramiento de petróleos extra-pesados de la Faja Petrolífera del Orinoco, 12/08/1993.

COMISIÓN BICAMERAL DE ENERGÍA Y MINAS, VENEZUELA, *Informe, Convenio de Asociación para Exploración a Riesgo de Nuevas Áreas y la Producción de Hidrocarburos bajo el Esquema de Ganancias Compartidas*, 21/06/1995.

COMISIÓN BICAMERAL DE ENERGÍA Y MINAS, VENEZUELA, *Informe que presenta la Comisión Bicameral de Energía y Minas del Congreso de la República, sobre el Marco de Condiciones que regirán el Convenio de Asociación para la Explotación, Transporte, Mejoramiento y Comercialización de Crudos Extra-Pesados del Área Cerro Negro, ubicada en la Faja Petrolífera del Orinoco, a celebrarse entre las Empresas Lagoven, S.A., Mobil Oil Corporation y Veba Oel AG*, 10/04/1997.

COMISIÓN BICAMERAL DE ENERGÍA Y MINAS, VENEZUELA, *Informe que presenta la Comisión Bicameral del Congreso de la República, sobre la Autorización del Marco de Condiciones del Convenio de Asociación entre las Empresas Corpoven, S.A., Atlantic Richfield Co., Phillips Petroleum Company y Texaco Inc. para la Explotación, Producción Mezcla, Industrialización, Transporte, Refinación, Mejoramiento y Comercialización de Crudos Extra-Pesados del Área Hamaca de la Faja Petrolífera del Orinoco*, 16/04/1997.

COMISIÓN PERMANENTE DE ENERGÍA Y MINAS CÁMARA DE DIPUTADOS, VENEZUELA, *Informe sobre los Contratos de la Industria Petrolera Nacional para Reactivar los Campos Inactivos o Marginales*, 11/1992.

CORONIL Fernando, *The Magical State*, University of Chicago, 1997.

CORPOVEN, *Convenios de Asociación para Exploración a Riesgo de Nuevas Áreas y la Producción de Hidrocarburos bajo el Esquema de Ganancias Compartidas*, 1995.

CORPOVEN, *Asociación entre las Empresas Corpoven, S.A. y Atlantic Richfield Co. para la Explotación y Mejoramiento de Crudo Extrapesado del Área Hamaca de la Faja del Orinoco*, 11/1996.

CORPOVEN, *Convenio de Asociación Hamaca (Ameriven)*, 09/07/1997.

CORTE SUPREMA DE JUSTICIA DE VENEZUELA, *Sentencia*, 23/04/1991.

CURTICE Arthur A. and HOOVER JR Herbert., *Mémorandum*, Caracas, 02/01/1943.

DE LA PLAZA Salvador, *Estructuras de Integración Nacional*, Caracas, Editorial Pensamiento Vivo, 1959.

DUQUE CORREDOR Román J., *El Derecho de la Nacionalización Petrolera*, Caracas, Editorial Jurídica Venezolana, 1978.

EGAÑA Manuel R., *Memoria del Ministerio de Fomento*, Caracas, 1939.

EGAÑA Manuel R., *Introducción a la Memoria del Ministerio de Fomento*, Caracas, 1941.

EGAÑA Manuel R. and PIZANI Rafael, "Informe que presentan al Presidente de la República, acerca del Proyecto de Ley de Petróleos", Caracas. Personal archives of Manuel R. Egaña, 06/01/1943.

EGAÑA Manuel R., *Tres Décadas de Producción Petrolera*, Caracas, Tipografía Americana, 1947.

EGAÑA Manuel R., "Instrucciones de los Ministerios de Relaciones Exteriores y de Fomento para los Doctores Edmundo Luongo Cabello, Luis E. Monsanto y Ezequiel Monsalve Casado, miembros de la Comisión Especial que envía Venezuela cerca de los países del Medio Oriente y el Egipto", Personal archives of Manuel R. Egaña, 1949.

EGAÑA Manuel R. and MAZA ZAVALA Domingo Felipe, *Nacionalización petrolera en Venezuela*, Caracas, Monte Ávila Editores, 1971.

EGAÑA Manuel R., *Venezuela y sus Minas*, Banco Central de Venezuela, Caracas, 1979.

El País, Venezuelan newspaper, Caracas, 28/12/1946.

El Nacional, Venezuelan newspaper, Caracas.

ESPINASA Ramón and MOMMER Bernard, "Venezuelan Oil Policy in the Long Run", in DORIAN James P. et FESHARAKI Fereidun (eds.), *International Issues in Energy Policy, Development and Economics*, Boulder Colorado, Westview Press, 1992.

FARMAN Farmaian and MANUCHER Mirza, *Blood and Oil: Memoirs of a Persian Prince*, New York, Random House, 1997.

Fortune, "Creole Petroleum: Business Embassy", USA, monthly publication, 02/1949.

Gaceta Oficial de Venezuela.

GARCÍA MACÍAS Carlos Luis and MOMMER Irama, *El régimen concesionario del petróleo venezolano : aspectos jurídicos, institucionales y políticos*, Amazon, 2018.

GONZÁLEZ RUFINO Miranda, *Estudios acerca del Régimen Legal del Petróleo en Venezuela*, Caracas, 1958.

Great Britain, High Court of Justice, *Folio 61*, London, 20/03/2008.

GUERRA José, *El Legado de Chávez*, Caracas, Editorial Libros Marcados, 2013.

HENDRYX Frank C., "A Sovereign Nation's Legal Ability to Make and Abide by a Petroleum Concession Contract", *Platt's Oilgram News*, 28/04/1959.

INTERNATIONAL CENTRE FOR SETTLEMENT OF INVESTMENT DISPUTES, *COP vs. BRV. Decision on Jurisdiction and Merits*, 2013.

INTERNATIONAL CENTRE FOR SETTLEMENT OF INVESTMENT DISPUTES, *XOM vs. BRV: Award*, 2014.

INTERNATIONAL CENTRE FOR SETTLEMENT OF INVESTMENT DISPUTES, *COP vs. BRV Decision on Jurisdiction and Merits, Georges Abi-Saab: Dissenting Decision*, 19/02/2015.

INTERNATIONAL CENTRE FOR SETTLEMENT OF INVESTMENT DISPUTES, *BRV vs. XOM: Annulment Decision*, 2017.

INTERNATIONAL CENTRE FOR SETTLEMENT OF INVESTMENT DISPUTES, *COP vs. BRV: Award*, 2019.

INTERNATIONAL CENTRE FOR SETTLEMENT OF INVESTMENT DISPUTES, *BRV vs. COP: Application for Annulment*, 08/07/2019.

INTERNATIONAL CHAMBER OF COMMERCE, *Case No. 1546/JRF/CA, XOM vs. PDVSA*, Paris, 23/12/2011.

INTERNATIONAL CHAMBER OF COMMERCE, *Case No.20549/ASM/JPA, COP vs. PDVSA*, Paris, 24/04/2018.

INTERNATIONAL ENERGY AGENCY, *Oil Market Report*, monthly publication.

INTERNATIONAL MONETARY FUND, *The Impact of Higher Oil Prices on the Global Economy*, Research Department, 08/12/2000.

JEVONS William Stanley, *The Coal Question*, London, Macmillan & Co., 1865.

KAUTSKY Karl, *Die Agrarfrage*, , J. H. W. Dietz, Stuttgart, 1899.

KIELMAS Maria, "Venezuela: Little Moves Ahead of an Explosion", *Petroleum Economist*, 11/1992.

LAGOVEN, *Proyecto Cristóbal Colón.* Anexos: "A": *Copia del Acta del Consejo de Ministros del 24 de mayo de 1991*; "B": *Circunstancias Pertinentes para la Asociación*; "C": *Condiciones Propuestas para el Convenio de Asociación a ser Consideradas por el Congreso de la República*; "D": *Proceso de Toma de Decisiones en la Sociedad a ser Creada*, 06/1991.

LAGOVEN, *Proyecto Cristóbal Colón.* Anexos: I. *Convenio de Asociación y Anexos*; II. *Condiciones Propuestas para el Convenio de Asociación a ser Consideradas por el Congreso de la República*; III. *Proceso de Toma de Decisiones en la Sociedad a ser Creada*; IV. *Acuerdo de Formación*; V. *Garantías*, 03/1993.

LAGOVEN, *Proyecto Cerro Negro*, 1996.

LAGOVEN, *Convenio de Asociación Cerro Negro*; *Anexo*: *Convenio de Reserva y Dedicación*, 28/10/1997.

LAMEDA Guaicaipuro, *Pronunciamiento ante el Proyecto de Ley de Hidrocarburos*, Caracas, 08/11/2001.

LICHTBLAU John, and SPRIGGS Dillard P., *The Oil Depletion Issue*, New York, Petroleum Industry Research Foundation, 1959.

LUONGO CABELLO Edmundo, Ministerio de Fomento, *El Aumento de los Impuestos Petroleros en el Proyecto de Ley de Hidrocarburos*, 1943, in VALLENILLA Luis, *Auge, Declinación y Porvenir del Petróleo Venezolano*, Caracas Editorial Tiempo Nuevo, 1973.

LUTFI Ashraf, *OPEC Oil*, Beirut, Middle East Research & Publishing Center, 1968.

MARAVEN, *Asociación entre las empresas Maraven y Conoco para la explotación y mejoramiento de crudo extrapesado de la Faja del Orinoco*, 04/1993.

MARAVEN, *Asociación entre las empresas Maraven, Total, Itochu y Marubeni para la explotación y mejoramiento de crudo extrapesado de la Faja del Orinoco*, 06/1993.

MARAVEN, *Convenio de Asociación Petrozuata,* 1995.

MARAVEN, *Convenio de Asociación Sincor,* 1997.

MARSHALL Alfred, *Principles of Economics*, London, Macmillan, 1890.

MARTÍNEZ Aníbal R., *Nuestro Petróleo. Defensa de un Recurso Agotable*, Genève, 1963.

MARTÍNEZ Aníbal R., *Recursos de Hidrocarburos de Venezuela*, Caracas, Edreca Editores, 1972.

MARX Karl, *Theorien über den Mehrwert*, vol. II, Marx Engels Werke Band 26.2, Berlin, 1967 [1861-1863].

MARX Karl, *Das Kapital*, vol. III, Marx Engels Werke Band 25, Berlin, 1966 [1882].

MAYOBRE José Antonio, *La Paridad del Bolívar*, Caracas, Banco Central de Venezuela, 1982.

MAZA ZAVALA Domingo Felipe, *Venezuela – una economía dependiente*, Caracas, Universidad Central de Venezuela, 1964.

MEJÍA ALARCÓN Pedro Esteban, *La industria del petróleo en Venezuela*, Caracas, Universidad Central de Venezuela, 1972.

MELCHER Dorothea, *La Organización de los Países Exportadores de Petróleo. La Prehistoria de su Fundación (1943 a 1960)*, to be published, 2010.

MIKDASHI Zuhayr, *A Financial Analysis of Middle Eastern Oil Concessions : 1901-1965*, New York, Praeger, 1966.

MIKDASHI Zuhayr, *The Community of Oil Exporting Countries*, London, Routledge, 1972.

MINISTERIO DE ENERGÍA Y MINAS DE VENEZUELA, *Convenio de Regalía con Bitor,* 11/11/1993.

MINISTERIO DE ENERGÍA Y MINAS DE VENEZUELA, *Convenio de Regalía con Bitor*, 08/1996.

MINISTERIO DE ENERGÍA Y MINAS DE VENEZUELA, *Convenio de Regalía Mejoramiento de crudos extrapesados*, 29/05/1998.

MINISTERIO DE ENERGÍA Y MINAS DE VENEZUELA, *La Desnacionalización del Petróleo Venezolano en los Años Noventa: II. Los Convenios Operativos*, Report addressed to the National Assembly, 08/2005.

MINISTERIO DE ENERGÍA Y MINAS DE VENEZUELA, *La Desnacionalización del Petróleo Venezolano en los Años Noventa: I. Internacionalización*, Report addressed to the National Assembly, 09/2005.

MINISTERIO DE ENERGÍA Y MINAS DE VENEZUELA, *La Desnacionalización del Petróleo Venezolano en los Años Noventa: III. Los Convenios de Asociación*, Report addressed to the National Assembly, 03/2006.

MIRABEAU Comte de, *Collection complète des travaux de M. Mirabeau l'aîné, à l'assemblée nationale*, vol. V, Paris, 1792.

MOMMER Bernard, *Die Bedeutung der Grundrente in der Petroleumproduktion. Analytische Geschichte ihrer Entwicklung von den Anfängen in den USA bis zur internationalen Ölkrise der Gegenwart*, doctoral thesis, Tübingen, 1977.

MOMMER Bernard, "Valores internacionales y los términos absolutos de intercambio del petróleo venezolano (1917-1977)", in FLICHMAN Guillermo, HAGEDOORN Leo and STROOM Jean (eds.), *Renta del suelo y economía internacional*, Amsterdam, CEDLA Incidentele Publicaties 19, 1981.

MOMMER Bernard, *Die Ölfrage*, Hamburg, Nomos Verlagsgesellschaft Baden-Baden, 1983.

MOMMER Bernard, "Oil Rent and Rent Capitalism: The Example of Venezuela", *Review*, Fernand Braudel Center of New York, vol. XIII, No. 4, 1990.

MOMMER Bernard, "¿Es posible una política petrolera no rentista?", *Revista del Banco Central de Venezuela*, vol. 4, No 3, Caracas, 1989.

MOMMER Bernard, "La Distribución de la Renta Petrolera – El Desarrollo del Capitalismo Rentístico Venezolano", in BELLO Omar and VALENCILLO Héctor (eds.), *La Economía Contemporánea de Venezuela*, vol. IV, Caracas, Banco Central de Venezuela, 1990.

MOMMER Bernard, "Integrating the Oil: A Structural Analysis of Petroleum in the Venezuelan Economy", *Latin American Perspectives*, Issue 90, vol. 23, No. 3, 1996.

MOMMER Bernard, *Private Landlord-Tenant Relationship in British Coal and American Oil*, Oxford Institute for Energy Studies, 1997.

MOMMER Bernard, "The New Governance of Venezuelan Oil", *Oxford Institute for Energy Studies*, 04/1998.

MOMMER Bernard, "Propiedad mineral y nacionalización de la industria petrolera mexicana", *Venezuela en Oxford – 25 años de la Cátedra Andrés Bello en el St. Antony's College de la Universidad de Oxford*, Banco Central de Venezuela, Caracas, 1999.

MOMMER Bernard, *Global Oil and the Nation State*, Oxford University Press, 2002.

MOMMER Bernard, "La Orimulsión: Verdades Científicas y Mentiras Políticas", *Interciencia*, vol. 29, No. 1, Caracas, 2004.

MOMMER Bernard, "Venezuela y la OPEP Cincuenta Años más tarde", Supplement of *Le Monde Diplomatique*, Venezuelan Edition, Year III, 03/2011.

MOMMER Dorothea, *Venezuela 1936-1948. Parteien und Gewerkschaften in einem Petroleumland*, doctoral thesis, Tübingen, 1977.

MOMMER Irama, "La Política Petrolera Venezolana hacia las Refinerías de las Antillas Neerlandesas: Una Perspectiva Histórica", in SERBIN Andrés (ed.), *Venezuela y las Relaciones Internacionales en la Cuenca del Caribe*, Caracas, Instituto Latinoamericano de Investigaciones Sociales, 1987.

Nacionalización del Petróleo en Venezuela. Tesis y Documentos Fundamentales, Caracas, Catalá / Centauro Editores, 1975.

ORGANIZATION OF THE PETROLEUM EXPORTING COUNTRIES, *Resolutions*.

PAGE H. W., "Profit Sharing Between Producing Countries and Oil Companies: A Reply", *The Economic Journal*, vol. LXX, 09/1960.

PARRA Alirio, Ministro de Energía y Minas, Venezuela, *Comunicación al Presidente del Congreso de la República,* 04/08/1992.

PELTZER Ernesto, *Ensayos sobre economía*, Caracas, Banco Central de Venezuela, 1965.

PÉREZ ALFONZO Juan Pablo, *Voto Salvado*, 1943, in VALLENILLA Luis, *Auge, Declinación y Porvenir del Petróleo Venezolano*, Caracas, Editorial Tiempo Nuevo, 1973.

PÉREZ ALFONZO Juan Pablo, *Petróleo, jugo de la tierra*, Caracas, Editorial Arte, 1961.

PÉREZ ALFONZO Juan Pablo, *El Pentágono Petrolero*, Caracas, Ediciones Revista Política, 1967.

PÉREZ ALFONZO Juan Pablo, *Rescatemos el poder tarifario*, Caracas, Editor Guarenas, 1970.

PÉREZ ALFONZO Juan Pablo, *Petróleo y Dependencia*, Caracas, Síntesis Dos Mil, 1971.

PÉREZ ALFONZO Juan Pablo, *Hundiéndonos en los Excrementos del Diablo*, Caracas, Editorial Lisbona, 1976.

PETRÓLEOS DE VENEZUELA, S.A., *Documento Marco de las Asociaciones Estratégicas para la Explotación y Mejoramiento de los Crudos de la Faja de Orinoco y otros Crudos Pesados y Extrapesados*, Junta Directiva, 18/02/1993.

PETRÓLEOS DE VENEZUELA S.A., *Proyecto Araguaney*, 15/03/1997.

PLATT'S OILGRAM, 28/04/1959.

POGUE Joseph E., "Oil in Venezuela", *Chase National Bank*, Junta Directiva, New York, 06/1949.

QUESNAY François, *Tableau Économique*, Versailles, 1758.

RATHBONE M. J., *The Financial Times*, London, 06/03/1958.

REVISTA DE DERECHO Y DE LEGISLACIÓN, Year XLIII, No. 515-518, Caracas, april-july 1954.

RICARDO David, *On the Principles of Political Economy and Taxation*, London, John Murray, 1817.

RIVERO Mirtha, *La Rebelión de los Náufragos*, Caracas, Editorial Alfa, 2010.

RODRÍGUEZ Gumersindo, *El Nuevo Modelo de Desarrollo Venezolano*, Caracas, Ediciones Corpoconsult, 1979.

RODRÍGUEZ ARAQUE Alí, *Gaceta del Congreso,* 03/08/1993.

RUPTURA, COMISIÓN IDEOLÓGICA DE (1975-1979), *El Imperialismo Petrolero y la Revolución Venezolana*, Vol. I: *Capital y Propiedad Territorial*, Ángel J. Márquez (ed.), Caracas 1975; Vol. II: *Las Ganancias Extraordinarias y la Soberanía Nacional*,

Ángel J. Márquez (ed.), Caracas 1977; Vol. III: *La OPEP y las Nacionalizaciones: La Renta Absoluta*, Ramón Rivero (ed.), Caracas 1979.

SCOTT Richard, *The History of the International Energy Agency – The First Twenty Years; Vol. II: Major Policies and Actions*, Paris, 1995.

SECRETARIAT OF THE ENERGY CHARTER TREATY, *El Tratado de la Carta de la Energía y Documentos Relacionados*, 1995.

BILATERAL INVESTMENT TREATY VENEZUELA-NETHERLANDS, *Agreement on Encouragement and Reciprocal Protection of Investments between the Kingdom of the Netherlands and the Republic of Venezuela*, 1993.

SECRETARIAT OF THE ENERGY CHARTER TREATY AND INTERNATIONAL ENERGY AGENCY, *Energy Investment*, 01/04/1998.

SILVA CALDERÓN Álvaro, *Gaceta del Congreso*, Sessions 28/07/1993, 04/08/1993 and 10/8/1993.

SMITH Adam, *An Inquiry into the Nature and Causes of the Wealth of Nations*, London, 1776.

SOCIÉTÉ DES NATIONS, *Journal officiel*, vol. 13/4, 1932-1933.

SOSA PIETRI Andrés, *Conferencia dictada por Andrés Sosa Pietri,* President of PDVSA, IV Venezuela Petroleum Congress, Caracas, 18/07/1990.

SOSA PIETRI Andrés, *Petróleo y Poder*, Caracas, Planeta Venezolana SA, 1993.

STOBAUGH Robert B., "The Evolution of Iranian Oil Policy 1925-1975", in LENCZOWSKI George (ed.), *Iran under the Pahlevi*, Hoover Institution Press, 1978.

STOCKING George W., *Middle East Oil*, Nashville, Vanderbilt University Press, 1970.

STONE Richard and STONE Giovanna, *National Income and Expenditure*, Cambridge, 1961.

UNITED NATION ORGANIZATION, *Resolutions*.

UNITED STATES SENATE, Select Committee on Small Business, *The International Petroleum Cartel*, Staff Report to the Federal Trade Commission, 1952.

USLAR PIETRI Arturo, "Sembrar el Petróleo", Venezuelan Newspaper *Ahora*, Caracas, 14/07/1936.

USLAR PIETRI Arturo, *Sumario de Economía Venezolana*, Caracas, Universidad Central de Venezuela, 1945.

USLAR PIETRI Arturo, *Venezuela en el petróleo*, Caracas, Urbina y Fuentes Editores Asociados, 1948.

USLAR PIETRI Arturo, *Venezuela, un país en transformación*, Caracas, Universidad Central de Venezuela, 1958.

USLAR PIETRI Arturo, *Petróleo de Vida o Muerte*, Caracas, Editorial Arte, 1966.

VALLENILLA Luis, *Auge, Declinación y Porvenir del Petróleo Venezolano*, Caracas, Editorial Tiempo Nuevo, 1973.

VALLENILLA Luis, *La Apertura Petrolera – un Peligroso Retorno al Pasado*, Caracas, Ediciones Porvenir, 1995.

WORLD PETROLEUM, vol. 30, 01/1959.

Table of contents